The Shortcut

An influential scientist in the field of artificial intelligence (AI) explains its fundamental concepts and how it is changing culture and society.

A particular form of AI is now embedded in our tech, our infrastructure, and our lives. How did it get there? Where and why should we be concerned? And what should we do now? *The Shortcut: Why Intelligent Machines Do Not Think Like Us* provides an accessible yet probing exposure of AI in its prevalent form today, proposing a new narrative to connect and make sense of events that have happened in the recent tumultuous past, and enabling us to think soberly about the road ahead.

This book is divided into ten carefully crafted and easily digestible chapters. Each chapter grapples with an important question for AI. Ranging from the scientific concepts that underpin the technology to wider implications for society, it develops a unified description using tools from different disciplines and avoiding unnecessary abstractions or words that end with -ism. The book uses real examples wherever possible, introducing the reader to the people who have created some of these technologies and to ideas shaping modern society that originate from the technical side of AI. It contains important practical advice about how we should approach AI in the future without promoting exaggerated hypes or fears.

Entertaining and disturbing but always thoughtful, *The Shortcut* confronts the hidden logic of AI while preserving a space for human dignity. It is essential reading for anyone with an interest in AI, the history of technology, and the history of ideas. General readers will come away much more informed about how AI really works today, and what we should do next.

Nello
Cristianini

The Shortcut
Why Intelligent Machines Do Not Think Like Us

CRC Press
Taylor & Francis Group
Boca Raton London New York

CRC Press is an imprint of the
Taylor & Francis Group, an **informa** business

Cover Image Credit: Shutterstock.com

First edition published 2023
by CRC Press
6000 Broken Sound Parkway NW, Suite 300, Boca Raton,
FL 33487–2742

and by CRC Press
4 Park Square, Milton Park, Abingdon, Oxon, OX14 4RN

CRC Press is an imprint of Taylor & Francis Group, LLC
© 2023 Nello Cristianini

Library of Congress Cataloging-in-Publication Data
Names: Cristianini, Nello, author.
Title: The shortcut : why intelligent machines do not think like us / Nello
   Cristianini.
Other titles: Why intelligent machines do not think like us
Description: First edition. | Boca Raton : CRC Press, 2023. | Includes
   bibliographical references and index.
Identifiers: LCCN 2022045908 (print) | LCCN 2022045909 (ebook) |
   ISBN 9781032371993 (hardback) | ISBN 9781032305097 (paperback) |
   ISBN 9781003335818 (ebook)
Subjects: LCSH: Artificial intelligence—Social aspects. | Intellect. | Civilization,
   Modern—21st century.
Classification: LCC Q335 .C75 2023 (print) | LCC Q335 (ebook) | DDC 006.3—
   dc23/eng/20221130
LC record available at https://lccn.loc.gov/2022045908
LC ebook record available at https://lccn.loc.gov/2022045909

ISBN: 978-1-032-37199-3 (hbk)
ISBN: 978-1-032-30509-7 (pbk)
ISBN: 978-1-003-33581-8 (ebk)

DOI: 10.1201/9781003335818

Typeset in Sabon
by Apex CoVantage LLC

"Dedicated to all my teachers"

Contents

ABOUT THE AUTHOR   ix
PROLOGUE   xi

1 The Search for Intelligence   1
2 The Shortcut   16
3 Finding Order in the World   38
4 Lady Lovelace Was Wrong   51
5 Unintended Behaviour   62
6 Microtargeting and Mass Persuasion   76
7 The Feedback Loop   97
8 The Glitch   118
9 Social Machines   128
10 Regulating, Not Unplugging   148

EPILOGUE   161
BIBLIOGRAPHY   165
INDEX   172

About the Author

Nello Cristianini has been an influential researcher in the field of machine learning and AI for over 20 years. He is Professor of Artificial Intelligence at the University of Bath, before that he has worked at the University of Bristol, the University of California (Davis), the University of London (Royal Holloway). For his work in machine learning, he has been a past recipient of the Royal Society Wolfson Research Merit Award, and of the ERC Advanced Grant. Cristianini has been the co-author of influential books in machine learning, as well as dozens of academic articles, published in journals that range from AI to the philosophy of science, from the digital humanities to natural language processing, and from sociology to biology. In 2017, he delivered the annual STOA lecture at the European Parliament on the topic of the social impact of AI, a theme that he is still actively investigating. Cristianini has a degree in Physics from the University of Trieste, an MSc in Computational Intelligence from the University of London, and a PhD from the University of Bristol.

PROLOGUE

At age fourteen, I spent a lot of time with my computer, which would not be strange except that this was 40 years ago, in a small town in Italy. At school, I had to study Latin and Greek, and that winter I was having trouble with Greek grammar, so my mother organised for an old priest of the nearby convent to give me some extra tutoring. Don Antonio was about 80 years old, and I could not understand all that he said, but at least I could walk there on my own.

   One day, he asked me why I had not finished my work and so I had to tell him about the computer. He immediately wanted to see it, as he had read so much about computers in the newspapers, so the next day he walked to my home. It took a while. When he saw my little Sinclair ZX80, he looked a bit disappointed, probably because it was so small. I turned on the big black and white TV that I used as a monitor, and after the usual five minutes of warming up, we started seeing the faint image of the prompt.

"Ask it when Alexander the Great was born".

I objected that the computer would not know how to answer, but he strongly insisted that I should at least try. I was not going to argue with him, so I just typed the question, in Italian where commands in Basic language were expected, and I obtained the only possible answer: SYNTAX ERROR.

Don Antonio started making his impatient expression, which I knew all too well, so I told him, "If you let me do some preparation, I can program it to answer this question", and he went to speak with my mother.

A few minutes later I called him, and this time, after typing the question, the computer printed on the screen the answer, "356 BC". This annoyed him even more. "Of course it knows it now, you just told it the answer".

I explained to him that computers do not know anything on their own, we need to tell them everything, but then they can remember it and maybe reason about it. But by then he had turned his back on the machine and added,

"This thing will never be better than us".

He sounded a bit relieved, as well as disappointed.

Was he right, though? Had we just established that our creations shall always be inferior to us? He certainly did get me thinking for quite some time.

It must have been the summer of 2012, 30 years later, when my children and their cousins were playing with an iPhone on the beach, not far from the same town where the old priest saw my new computer. They showed me Siri, which I had read about but had never seen, and invited me to ask it any question.

"When was Alexander the Great born?" The question came unexpectedly out of some forgotten corner of my memory.

"Alexander the Great was born in Pella in 356 BC" the phone answered.

1

THE SEARCH FOR INTELLIGENCE

Intelligence is not about being human, but about being able to behave effectively in new situations. This ability does not require a brain: we can find it also in plants, ant colonies and software. Different agents can display this ability to different extents in different tasks: there is not a single general way to be intelligent, nor a secret formula or a single test. It is misleading to attribute human qualities to all intelligent agents, and when reflecting on the intelligences we encounter in our browsers we would do better by comparing them with the snails and herbs in our garden, rather than with us.

In 1972 and 1973, NASA launched two Pioneer spacecraft destined to ultimately leave the solar system. Along with their instruments, they also carried a "message from mankind" to any form of alien intelligence that might stumble upon them,

DOI: 10.1201/9781003335818-1

prepared by the astronomer Carl Sagan. It was a metal plaque with a drawing of two human beings, a map of the solar system, and the diagram of an atom of hydrogen, as well as other symbols such as a map of 14 pulsars. The idea behind the message was that it should enable an alien intelligence to identify its origin and authors, though to this day we are still to receive any response.

A few years later, Sagan—a long-time advocate of the search for extraterrestrial intelligence—gave a lecture at the Royal Institution in London, where he presented his newest effort on the same subject. This did not require aliens physically approaching an artefact from Earth, but just receiving a radio signal: a particular sequence of 31^3 bits that—once folded into a 31×31×31 cube—represents a stylised image of a formaldehyde molecule. This in turn would suggest to aliens that they should start listening on the radio frequency specific to that molecule. His argument: any aliens, wherever they might be, will still have evolved in the same universe and be subject to the same physical laws, so they would be able to understand these concepts.

To test this assumption, Sagan gave that message to four of his PhD students in physics, with no further instructions, and they managed to decode it, at least partially.

In the name of reproducibility, recently I have tried to repeat that experiment by showing both the plaque and the sequence to my cat, who not only has evolved in the same universe but also shares with me most of her evolutionary history, and even grew up in the very same house where I live. At the time of writing, I have yet to receive a response.

Yet the same cat can very effectively learn, plan, reason, and even communicate with other cats and humans. She knows what she wants, and she can find a way to get it, and if needed she can outwit not just a mouse or a dog but also me.

Was Sagan's plan missing something?

MANAGING EXPECTATIONS

How do we know if something is intelligent? If Sagan's messages reached most parts of the Earth today, or any part of Earth in most of its history, they would not have found anyone interested. Yet intelligence existed on this planet long before the first human appeared or the first word was spoken: predators hunted in packs, birds escaped predators, mice outwitted birds in order to eat their eggs, and so on. Even ant colonies made complex and well-informed decisions about the ideal place to nest. Are those not expressions of intelligence?

This question is important, since if we do not know what we are looking for we might not recognise it when we see it. One of the main challenges when thinking about any other forms of intelligence is precisely to imagine something fundamentally different from us.

Carl Sagan was not alone in looking at the question in an anthropocentric way: when some of the early pioneers of artificial intelligence (AI) wanted to demonstrate their objectives in a prototype, they created a software tool called the Logic Theorist, which was an algorithm for proving theorems inspired by George Polya's influential mathematics book *How to Solve It*. The same group, which included Nobel prize winner Herbert Simon, went on to develop the General Problem Solver, which they presented as a form of "general" intelligence, capable of manipulating symbols in order to reason from axioms to theorems. The message was clear: behind all expressions of intelligence there is the same elusive ingredient, regardless of the specific domain in which it is applied, and this has something to do with formal reasoning about symbols. The names of their prototypes tell us a great deal.

Fortunately, these were not the only attempts to capture the notion of "intelligence", but they do illustrate one of the challenges of this program: just like mathematicians may regard mathematics as the pinnacle of intelligence, so humans in general tend to regard "humanity" in the same way. The demarcation between intelligent and non-intelligent behaviour has often been drawn so as to separate humans from other animals, typically in a way involving the use of language, tools, complex planning, or empathy. To this day, many tests of human intelligence would not be suitable for other animals, being based on recognising visual patterns and manipulating language. The same is true for machines: Turing's Test, for decades the only operational definition of machine intelligence, involves a computer fooling human judges into thinking it is human during a conversation.

This anthropocentric bias in how we imagine intelligence makes it more difficult for us to imagine and investigate "alien" forms of intelligence. By this I do not mean extraterrestrial beings, but just anything that is "other" than ourselves. How about squids, plants, or ant colonies? How about machinery? In this book we will consider them as some type of "alien intelligence"; empathy with the alien is very hard indeed, if not impossible, but it is just this effort that helps us to avoid the risk of imposing our own expectations on very different types of intelligent systems. One of the most difficult questions in the future of AI will involve precisely the possibility that our machines can know things that we ourselves cannot comprehend.

DEFINING INTELLIGENCE

To keep things simple, we will define intelligence in terms of the behaviour of an agent, which is any system able to act in its environment and use sensory information to make its decisions. We are specifically interested in autonomous agents, that is, agents which make their own decisions internally, and environments that can be partly affected by the actions of the agent.

Imagine a hungry and timid slug that has to decide which way to crawl after being born in your garden. We can think of its behaviour as the way it responds to a stimulus: this can include involuntary reflexes, learnt reactions, and—for higher animals—calculated actions too. These responses imply some form of *anticipation*, such as when a rabbit starts running at the sight of a fox, or the fox salivates at the sight of the rabbit. Any changes in behaviour due to experience are what we call "learning".

We will informally define intelligence of such agents as *"the ability to behave effectively in a variety of situations, including new ones"*, a definition that has been repeatedly found in different domains: for example, in economics, rational agents are defined as "aiming to maximise their utility" and cybernetics studies agents capable of "goal-seeking behaviour". In 1991, American engineer James Albus defined it as *"the ability of a system to act appropriately in an uncertain environment, where appropriate action is that which increases the probability of success"*. The definition was so simple that it stuck.

Importantly, we do not assume that there should be a single "quality" that makes agents intelligent, just like there is probably not a single quality that makes an athlete "fit". Different agents operating in different environments might just develop different tricks to behave effectively in unseen situations. In fact, it is easier to imagine that evolution produced a mixed bag of tricks, each of them adding some new capability, even within a single organism.

Certainly, we do not need to assume that the agent has a brain, language, or consciousness. But we do need to assume that the agent has goals, so that an effective action is one that increases its chances of achieving them. The idea of *purpose-driven* (or *purposeful*) behaviour is the cornerstone of *cybernetics*, an early science of intelligent agents. For example, in the game of chess, the environment is represented both by the rules and by the opponent, who responds to the agent's moves, and the purpose of the agent is to win a match by forcing a checkmate. In evolutionary biology, it is assumed that the ultimate goal of any organism is the survival of its genes.

Purposeful behaviour is also called *"teleological"* in the philosophy of science, from the Greek word "telos" which Aristotle used to define the ultimate purpose, or spontaneous direction of motion, of an entity. An agent that consistently moves towards a specific goal, whatever that is, can be considered as "teleological". In biology, this goal has to do with survival or reproduction, but other agents can have very different goals, such as scoring points in a video game or maximising profit in an online shop.

Notice that a consequence of this is that an agent needs a "body", that is, a way to interact with its environment, and it does not make much sense to consider 'disembodied' intelligence (with all due respect to Descartes). However, we do not require either bodies or environments to be physical, as we will see later in the case of digital agents who operate on the web. The body is just whatever allows the agent to affect, and be affected by, its environment.

If intelligence is just the ability to pursue one's objectives in a variety of different situations, including situations never experienced before, then there does not need to be a single elusive quality to be discovered, and we are free to analyse intelligent systems by focusing on whatever "tricks" they can deploy to effectively steer towards their goals in their respective environments.

The interaction between the environment and the goals of an agent is so tight, and the boundary between them so blurred, that often we will talk about "*task-environments*" to combine the two. For example, "finding glucose molecules in a Petri dish" or "selling books to online customers" are both goals to be reached within a specific environment, which dictates the basic rules, respectively, for a bacterium and an algorithm. For a human being, obtaining a high score in the game of Pac-Man is also a combination of task and environment. Sometimes, we will refer to this combination solely as "environment", assuming that the goals are fixed and clear, while environments can vary and so must be specified. Being able to succeed in a range of environments is what makes an agent robust, or general, or indeed intelligent. The space of viable environments for an agent can vary in many dimensions: a bird might have to forage for food in the same garden on different days, or in different seasons, or in different gardens, or even on two continents far apart. Clearly, different skills may be needed to move along this range.

For some philosophers, intelligence is about being able to solve different instances of certain natural puzzles while for others, it is all about being able to learn new types of puzzles; but actually, there is a spectrum between those two definitions. We will be satisfied with noticing that intelligent agents may need to be robust to changes across a range of environments and situations: a more specialised agent can cope well with task-environments that are more similar to each other, and a "broader" one can adapt to a larger diversity of task-environments. The first could be considered more of a specialist, the second more of a generalist, but it is always a matter of degree.

A REGULAR WORLD

We are implicitly assuming that it should be possible for an agent to behave effectively in a given task-environment, but this is not an obvious condition.

Consider a bird intent on increasing its blood sugar levels by pecking on the right type of berries: this is an opportunity to make good use of sensory information and prior experience so as to make the best decisions (and is just the type of situation that makes evolution of these very capabilities possible). But any such behaviour would be based on the implicit assumption that similar-looking berries will have similar taste and similar nutritional value, that the bird can sense the differences in colour and shape that correlate with nutritional content, and that it has the capability to act accordingly, namely, to execute its decisions. If the sugar content of a berry was completely independent of its colour, or the bird could not tell colours apart, or did not have the dexterity to pick and choose which berries to eat, then there would not be a benefit in being "intelligent"—nor evolutionary pressure to refine that ability.

A goal-oriented behaviour only makes sense in environments that are—at least in part—controllable and observable, or—using again the words of James Albus—*"where appropriate action increases the chances of success"*.

Only in a regular environment can an agent anticipate the future, and only there does it make sense for it to remember the correct response to any situation, which we could imagine as an ideal table, be it innate or learnt from experience.

An even stronger assumption, which lies behind learning and generalisation, is that similar actions will result in similar effects. It is this type of regularity that allows an agent to face an infinite variety of situations with finite knowledge, including situations never encountered before, and sometimes even new environments altogether.

These conditions cannot exist in a completely random environment, so at the very least we need to assume that an intelligent agent exists in a non-random—or regular—world. Even weak statistical regularities may be sufficient to confer some advantage to the agent, and discovering these regularities from experience is one of the ways in which intelligent behaviour is generated. Finding some order in the world is a requisite for purposeful behaviour, and we will discuss it in Chapter 3.

Evolution can take place after making some implicit assumptions: that the environment is regular and that there is some specific type of regularity that matches the organism's design. Given two different bacteria living in the same puddle of water, even a small difference in how they exploit environmental cues can translate into shorter replication times, and eventually exponential differences in growth. The organisms that we observe in nature are those that implicitly contain the best assumptions about the specific regularities that are present in their environment.

THE BAG OF TRICKS

Evolution is a pragmatist: so long as a behaviour leads the agent towards its ultimate goal, it does not matter how that behaviour is generated. It should not be surprising, then, if we find that similar behaviours in different species are achieved through different mechanisms: rather than a single quality that we can call "intelligence", behind intelligent behaviour we should expect to find a bag of tricks. Some of those, however, are so useful and distinct that they have their own name: *reflexes*, *planning*, *reasoning*, *learning*, and so on.

As mentioned above, a useful way, albeit an idealised one, to think about the behaviour of an agent is to imagine a table that lists every possible stimulus or situation and the action to take in response to each, even though this pairing, in reality, is performed by some mechanism and changes over time. This concept allows us to imagine reflexes as fixed pairings of stimulus with response, and learning as a change of these connections due to experience.

The way this table is actually implemented can vary; for example, for any given stimulus, a software agent would typically evaluate all possible responses based on their expected utility, and pick the most promising one, so that the programmer does not need to enumerate all possible inputs and corresponding outputs in advance. This technique could be described as a simple form of reasoning or planning and might also include a model of the environment; imagine an agent recommending different books to each customer, based on a rough description of both book and customer. A bacterium instead would choose its responses to the environment in an entirely different way that is based on a pathway of chemical reactions.

The question of how a biological agent—in this case, endowed with a brain—can do this was at the centre of an influential book with a memorable subtitle. Written in 1949 by the founder of neuropsychology, Donald Hebb, it was entitled *The*

Organization of Behavior and had this tagline written on its dust jacket: "Stimulus and Response, and what occurs in the brain in the interval between them". Indeed, what happens between them is just what makes psychology and AI such exciting disciplines: reflexes, planning, reasoning, and learning are all parts of the repertoire of tricks that can help an agent pursue its goals in a complex and changing environment. By interacting together, they contribute to making its behaviour "intelligent".

We will see in Chapter 3 how predictions can sometimes be made on the basis of patterns discovered in the data of past experience. These patterns could involve—for online shop agents—the segmentation of users and products into "types" whose behaviour is similar, an abstraction that allows for generalisation and compression of experience. Compressing large behaviour tables into simple formulas, or mechanisms, is a standard way for software agents to learn, avoiding exhaustive enumeration of all possible situations.

Whether they are based on statistical patterns, or on neural pathways, or on chemical reactions, the decisions made by intelligent agents do not need to involve sentience nor language, nor a higher appreciation of art and science. Humans do have all that, but they seem to be the exception rather than the rule.

ALIEN INTELLIGENCES

It was not just Carl Sagan who imagined intelligence in rather anthropocentric terms; while warning against this "prejudice" in 1948, Alan Turing, the pioneer of this field, also seemed to fall victim to the same temptation.

In a report entitled "Intelligent Machinery" which he never published, he *"propose[d] to investigate the question as to whether it is possible for machinery to show intelligent behaviour"* and suggested that the attribution of intelligence may be a subjective judgement; in other words, we might be tempted to attribute the same behaviour to "intelligence" when performed by a human and not when performed by a machine. To illustrate this point, he described a hypothetical test where a human judge is playing chess against an algorithm and against a human, and finding it difficult to tell which one is which. A few years later, this scheme formed the basis for the celebrated Turing Test, published in 1950, which only changed the type of game used for the interaction (in the later version, the chess match had been replaced by a conversation).

To be fair, Turing never implied in the report that being indistinguishable from a human is necessary for intelligence, just that it would be sufficient. But by using people as the paragon of intelligence, he did not seem to consider the many other forms of intelligence that exist on this planet, and in this he might have misled many researchers that followed. A few years later, computing pioneer, John McCarthy would coin the phrase "artificial intelligence" in the funding proposal for a workshop held in 1956, and in that very document he defined it as *"making a machine behave in ways that would be called intelligent if a human were so behaving"*.

Of the many myths that have been confusing the discussion on machine intelligence, one is particularly insidious: that humans are endowed with some sort of universal intelligence, so that there are no mental abilities in nature—and perhaps even in theory—that are beyond ours. This often goes along with the myth of a presumed superiority of the human brain compared with virtually any other product of evolution.

But there is no reason to believe that either of these myths is true. Our mental abilities are indeed astonishing, and it is very easy to come up with tasks that only we can complete; just look at any of the IQ (intelligence quotient) tests that we created. On the other hand, it is also equally easy to come up with tasks that we cannot complete, while other agents can. In other words, we are not the best at every cognitive task. Our mind is limited to a core set of assumptions about our task-environment, which make it rather specialised; we think in terms of objects, agents, basic geometry, and intuitive causality. Give us a task that violates those expectations, and we will be hopeless; what to do if the causes of a phenomenon simply do not exist, or are remote or indirect or distributed? We still struggle to comprehend the sub-nuclear world, where objects are replaced by waves and have no definite position, and where all our macroscopic metaphors fail.

We cannot read the barcode at the back of this book, or any QR codes, and we cannot detect a variety of patterns in datasets, nor even perform computations that require too much working memory. Yet all of these cognitive tasks are routinely performed by machines. We also see apes performing visual-memory tasks better than we do, and other animals displaying faster reflexes, sharper senses, and more capable memories, but each time we dismiss those examples as "not real intelligence".

An "alien" intelligence is an intelligence that is entirely different from ours, and therefore one we may be tempted to ignore; that of ant colonies, of machines, or of deep-sea creatures.

For example, how does an octopus think? It seems that some of its decisions are made autonomously by its tentacles, while others are made centrally, and it seems that changing its colour to match the environment or to signal emotion is an important behaviour. How do we relate to that alien, whose latest common ancestor with us existed 700 million years ago?

Yet those creatures are probably the best place to test our theories on alien intelligence, and one of the first questions that come to mind is: if humans are not the paragon of all intelligences, and even less their pinnacle, could there be other forms of intelligence that are "superior" to ours?

The problem with this question is that it implies a one-dimensional quantity (such as height or weight) that can be compared between two agents, asking which of the two will score higher. But the capabilities that we consider as expressions of "intelligence" seem to belong to multiple dimensions, so that it does seem more reasonable to expect different agents to be better at some tasks and worse at others.

It is this fallacy that leads some to imagine the existence of a "general purpose intelligence", which is often meant as a "universal" capability. There are two problems with this idea: the first is that there are different dimensions to intelligence, much like there are different ways for an athlete to be "fit"; is a marathon runner fitter than a swimmer or a cyclist? And the second one is that being general—or robust—is part of the very definition of an intelligent agent, since we ask for its behaviour to be effective under a range of situations.

We may find many intelligences that are "superhuman" for various specific tasks, both among animals and among machines, but we might be unable to find a single one that is "superhuman" at every task. There are software agents that can outperform every human at a range of board games and video games, and one day we might see them increasing the range of tasks where they excel. They may well become better than us at most tasks that we value. This still would not make them "universal", though at that point we might want to start worrying.

As we move away from an anthropocentric view of intelligence, we should not expect sentience or consciousness in our machines, the spontaneous emergence of a language that is translatable into ours, and emotional traits such as envy or fear that might affect our relation to these agents. All these are just projections of our incapability to imagine truly alien intelligences. Are we more likely to encounter "Promethean defiance" or "utter indifference" in a garden slug or a spam filter?

These expectations have the same roots as our older depictions of natural evolution as a ladder, with tadpoles on one side and human beings on the other, which implies that we are the natural destination of evolutionary processes. Just as we are not the pinnacle of evolution, so language and art are not the pinnacles of intelligence.

What we should expect of our intelligent artefacts is rather a relentless pursuit of simple goals, indifferent to any consequences for us, perhaps empowered by a capability to learn and improve performance with experience, even adapting to our possible countermeasures. This is—for example—what an agent does, when charged with increasing user engagement, by inducing more users to click. We call these agents "recommender systems" and we will use them as a running example, as they are today ubiquitous. There is no use in trying to have a conversation with them, for they are, after all, just very complex autonomous devices with no homunculus within them.

Ultimately, an algorithm is just a recipe. If we do not understand this and keep imagining the aliens of Carl Sagan or the sentient computers of Stanley Kubrick, we will not be able to develop the cultural antibodies that we will need to coexist with our creatures.

A COPERNICAN STEP

So, what is intelligence then? We may never find a demarcation between intelligent and non-intelligent agents that satisfies everyone, and maybe we do not need one, at least not urgently. Most likely our understanding of that concept will mature as we keep on building machines that display different types of intelligence.

One thing seems certain: Carl Sagan's clever messages intended for extraterrestrial intelligences would be lost on every type of intelligent agent on this planet, except for one. Rather than looking in deep space, he could have looked in our backyards: slugs relentlessly pursuing the pot of herbs, no matter how far you move it, plants constantly jockeying for a sunny position, and ant colonies pooling information together to make collective decisions about foraging and nesting. It is this type of intelligence that most closely resembles what we are building today in AI and are called to regulate by law. It would not be helpful to attempt to regulate a higher form of sentience that does not exist.

Most living things show some form of autonomous and purposeful behaviour, making a constant stream of informed decisions based on their innate goals and sensory information. They have limited control of their environment, but just enough that they can benefit from guessing, learning, and—in a few cases—reasoning and planning.

I am reasonably certain that the universe is full of intelligent entities, none of which has any interest in poetry or prime numbers, while they all exploit the regularities that they find in their environments to better pursue their goals. Recognising this as intelligent behaviour is part of letting go of the delusion that we, the humans, are the paragon of all things intelligent, a delusion that is holding back our understanding of the world. We should invoke one more time the old Copernican principle, according to which our impression of having a special place in the universe is just an effect of perspective, as it was for cosmology and biology.

That step would have the consequence of freeing our thinking, not only allowing us to recognise the intelligences that we encounter in our web browser whenever we use a recommender engine, but also to reset what we can expect from those creatures of ours. That step, more than anything else, will be essential if we are to learn how to coexist with them.

2

THE SHORTCUT

After attempting for years to capture and implement the elusive quality that makes things intelligent, researchers settled on studying and generating "purposeful behaviour" by various means. This resulted in a class of autonomous agents that can learn and behave effectively in a variety of new situations, by exploiting statistical patterns in their environment, which eliminated the need for explicit rules of behaviour, but introduced the need for large amounts of training data. This shortcut produced a new paradigm for the field, which is based on machine learning, and incorporates new expectations, new tools, and new success stories. Recommender systems are better representatives of AI agents than theorem provers. The new language of Artificial Intelligence is that of probability and mathematical optimization, no longer that of logic and formal reasoning.

"EVERY TIME I FIRE A LINGUIST"

"Whenever I fire a linguist our system performance improves". By 1988, Frederick Jelinek's team at IBM had solved enough

DOI: 10.1201/9781003335818-2

stubborn problems in speech recognition and machine translation to allow him to pick some bones at a conference speech.

An information theorist by training, from the 1970s he had pursued a radical approach to processing natural language, focusing on two apparently different problems that shared the same technical obstacle at their core. Due to various sources of ambiguity, both speech recognition and machine translation can generate a large number of possible outputs, most of which are nonsensical and can be easily ignored by a human; the problem is to tell machines which messages are meaningless, so as to keep only the plausible results.

Imagine that your speech recognition algorithm produced various possible transcriptions of your dictation, most of which look something like this:

"A a cats white shall called the the the girl and his mother are so is"

Even with a rudimentary knowledge of English we can flag this sentence as ill-formed, and if pressed we might come up with some reasons for it: articles should not follow other articles, adjectives should precede nouns, number and gender need to match between pronouns and nouns, and so do tenses of verbs. This is the path that most researchers had taken: using grammatical rules to define which sentences are "viable" as transcriptions or translations. The problem was that the rules never seemed to be enough, with each rule requiring another one and exceptions piling up. Grammars grew larger and the results remained weak.

This is when Frederick Jelinek joined IBM's Continuous Speech Recognition group in Yorktown Heights, NY, tasked with solving just this problem. His approach was not to incrementally build on the work of others.

Due to his training in information theory, his expertise was the reconstruction of messages that have been corrupted by some "noise" along the way, and he knew that just a coarse statistical analysis could reveal when a message had been so corrupted and even help us repair it. This is because every natural language has very stable statistical patterns: the probability of certain letter transitions remains constant across authors and domains, and so does that of certain words or phrases following each other. How many sentences have four verbs in a row, and in how many ways can you really complete the phrase "*once in a blue . . .*"?

Using these statistical regularities, Jelinek developed an approximate measure of how probable a given sentence (even one never seen before) is to occur in "natural" settings, and—even if rather crude—this could be used to automatically eliminate improbable translations, transcriptions, or sentence completions. As it turned out, this is often sufficient to help the algorithm find the most plausible candidates.

To make things simple, let us consider an easier example: the spell checker or auto-completion facility that we can find in all of our digital devices. Given a text that has been corrupted by typing mistakes, it is possible to correct the errors simply by substituting the words that have never been seen before (potential errors) with similar words that are also very frequent, or that are frequent in a similar context. This approach does not know grammar nor does it require an understanding of the topic of the text; all that is needed is statistical information of general type that can be obtained by analysing a large corpus of documents. These collections contain at least tens of thousands of documents and millions of words. From those, one needs to extract not just a list of all words and their frequencies, but also all pairs and triplets of words, which allow us to use context as a way to suggest a word; this means keeping track of millions of phrases and their frequencies.

The idea of eliminating grammar and replacing it with millions of parameters estimated from data went counter to mainstream wisdom, but this was the substance of Jelinek's intuition. He was used to doing things his way: a child migrant from

Czechoslovakia settled in the United States after World War II, he had studied information theory when his parents wanted him to become a lawyer, and then married a dissident filmmaker from Prague.

Those statistical patterns in language proved sufficient for the purpose of proposing the most plausible transcription of speech and translation of text, just as they are also used to propose the most plausible replacement of a mistyped word. They are so general that you could use the same method to process a language that you do not even speak, from Albanian to Zulu.

The catch is that these "patterns" are represented by massive tables, ideally listing the probabilities of millions of possible word combinations, which can only be learnt from millions of documents. But isn't that what computers are for? By the end of the 1980s, Jelinek's systems were the first ever to produce viable speech transcription and language translation, and in neither case did the algorithm have the faintest idea of the meaning of the words being manipulated.

The entire operation was statistical in nature and replaced the need for detailed models of natural language with a new need for large quantities of training data. But the idea was not just limited to language processing. This trade-off first faced by Jelinek foreshadowed later developments in other areas of AI; we can replace theoretical understanding with statistical patterns in other domains too, and we can obtain these patterns from massive sets of data.

When you can guess the next word in a sentence, guessing the next item in a shopping basket is not too far, and when you can do that, guessing the next movie that someone will want to watch is not that big a step either. The idea is the same; analyse samples of human behaviour to make statistical predictions in domains for which there is no practical theory.

Would these predictions always be guaranteed to work? Of course not. To use a phrase that would later become popular in the theory of machine learning, predictions are "*probably approximately correct"* at best, but that is all that is needed in many practical cases. And this is a crucial consideration for how the field developed. By the 1990s, Frederick Jelinek's radical idea would change not just the way we processed natural language, but also what we expected of our AIs. That was just the beginning.

GOOD OLD-FASHIONED AI

"The study is to proceed on the basis of the conjecture that every aspect of learning or any other feature of intelligence can in principle be so precisely described that a machine can be made to simulate it". John McCarthy was tired of "flying a false-flag", that is, of funding his research under the guise of information theory or automata studies; by 1956, he had decided to call things by the name he wanted, so he coined the phrase "artificial intelligence".

The first document containing that expression is the proposal for an event that would become legendary: the 1956 Dartmouth Workshop, the gathering of most researchers in the new field of machine intelligence. That document also contains the fundamental tenet of those days, that any feature of intelligence was so amenable to precise description as to be implementable in a machine.

This was not an unreasonable assumption; after all, if we want to calculate the trajectory of a projectile, we first need to understand mechanics, and if we want to design an aeroplane wing, we need to understand aerodynamics. This is how engineering has traditionally worked, and this is also why linguists were initially recruited to help in the design of language processing computers, by producing grammatical rules of language.

Appealing as it was, this obvious approach mostly highlighted how little we actually know of apparently simple phenomena such as vision and language; machines were able to prove theorems well before they were able to recognise cats.

These limitations became most visible in the 1980s, when competition from Japan stimulated massive investment in Europe and the United States in what was called "fifth-generation computer systems", which were essentially AI systems designed to be able to reason explicitly about a problem thanks to a "knowledge base" and logical inference rules. As those early AI systems were specialised in specific domains, their commercial name was "expert systems" and they were developed for all sorts of application areas: chemistry, medicine, oil fields. Knowledge about these domains of expertise was represented explicitly in symbols readable by humans and manipulated using logic.

The great evangelist of expert systems in the USA was Ed Feigenbaum, a famous computer scientist who repeated that intelligent machines cannot operate in the real world without a significant body of declarative knowledge to use for their logical reasoning and eventual decisions, whether these involve manipulating language, controlling a robot, or interpreting images. He called this method, which relied on human-readable facts and rules, the "knowledge-based paradigm" of AI; others called it "symbolic AI" but philosopher John Haugeland, in 1985, called it simply "Good Old-Fashioned AI", and the name caught on.

By the mid-1980s, expert systems were ubiquitous and synonymous with all AI; the main activity in the field consisted in casting practical problems into this framework, developing knowledge bases containing declarative statements, and heuristics to help combine and manipulate that knowledge. Imagine listing thousands of statements such as "wine contains alcohol" and "alcohol can be dangerous" and asking a computer to connect the dots. The expectations were high, fuelled by public investment, media coverage, and grand promises. Dedicated hardware was developed to speed up the computations of expert systems, such as the successful LISP machines.

Newspaper headlines from those days include optimistic statements such as "Human computer is coming of age" and "Smart Machines Get Smarter", and the documents of the US-based "Strategic Computing Initiative" list among their goals computer systems that could *"hear, speak and think like a human"*. A 1984 article in the New York Times says,

> *An international race for computer supremacy is under way (. . .) a radically new "fifth generation" or "artificial intelligence computer", that would have some reasoning power, would be incredibly simple for the average person to operate and would solve a whole range of problems.*

Hype and investment fed into each other, ultimately giving this idea the best chance of success one could hope for. And yet most promises remained unkept due to technical obstacles.

In the case of translation or speech recognition, the tasks faced at the same time by Frederick Jelinek with radically different techniques, this knowledge-based approach would have required parsers, morphological analysers, dictionaries, and grammars, each of those written in the form of handcrafted rules, which never seemed to be enough. The obvious answer for many was just to add more linguists.

The same was seen in all other areas where this kind of system was developed for the real world and a new job description even emerged: knowledge engineer, that is, someone capable of producing the rules needed by the machine in order to function. As researchers discovered how uncertain and ambiguous the real world actually is, their reaction was to produce ever more complex sets of rules and exceptions. In the end, expert systems turned out to be too brittle for real-world ambiguity and too expensive to maintain. There just did not seem to be clear logical theories for many domains of application.

After years of increasing hype and funding, AI conferences of that period reached their largest size, with researchers being joined by journalists, industrial recruiters, and salespeople. It was a time of optimism, and the golden age of symbolic AI and expert systems lasted for most of the 1980s, hitting its high-water mark in 1987. In the case of the International Joint Conference on AI (IJCAI-1987), held in Milan that year, this was more than a metaphor, since the conference hall was flooded by a broken water pipe, causing a suspension of the proceedings.

The first sign that something was changing was the collapse of the LISP machines market, the dedicated hardware for symbolic AI. From there, investment started to dry up, due to the limited usability of those brittle and expensive systems, which ultimately failed to keep up with the hype that preceded them.

A March 1988 article in the New York Times entitled "Setbacks for Artificial Intelligence" describes a rapidly changing mood of investors and markets about AI and explains it simply in terms of broken promises; "(. . .) *the setback stems from the failure of artificial intelligence to quickly live up to its promise of making machines that can understand English, recognise objects or reason like a human expert"*.

By that year, the massive (public and private) investment in symbolic rule-based AI had started drying up, leaving behind some success stories, but nothing close to the extravagant promises of the previous years.

That was the start of a period of downshifting and reflection for this young research field, and 1988 was also the same year in which the "Workshop on the Evaluation of Natural Language Processing Systems" was held at the Wayne Hotel in Wayne, PA, where Jelinek quipped about firing linguists, and when he published the description of the first entirely statistical system for machine translation.

SEASONS: BOOM, BUST, AND WINTER

The field of AI has seen various boom-and-bust cycles, some smaller and some larger, where a stage of inflated expectations is followed by a period of reduced investment. This volatility is not unusual in new technologies, but AI does seem to have made a habit of it, embracing its fallow periods and even naming them "winters". Although the term "AI winter" was originally coined in analogy with "nuclear winter", which was popular during the Cold War, it is now often used to describe a less "dramatic" and even cyclic scenario of reduced funding. The years following the collapse of the LISP machines market are now known as "the second AI winter".

These wintery periods can be healthy times, both for the reduced media coverage and for the reduced temptation to over-promise, making the sector less appealing to mavericks, and creating a market for new theoretical ideas that can thrive in a low-funding environment. The second winter started in the late 1980s and created space for ideas that had been incubating for years in specialist niches, among which were statistical methods for pattern recognition and early work on neural networks. These more mathematical and less ambitious methods were developed for industrial applications, without the pretence of solving anything near general intelligence: their concerns included the recognition of simple images such as handwritten digits, the retrieval of documents from corporate databases, the mining of datasets of commercial transactions, and so on. Jelinek's work on translation is just an example of how statistics was applied to solving these practical tasks.

Importantly, these years also saw increased academic interest in the very idea of "machine learning", initially by separate communities that developed independently. New methods were discovered for training neural networks, first proposed as models of nervous systems; for the inference of decision trees as a way to analyse data; and also for learning logical rules from examples, as part of an attempt to remedy the shortcomings of expert systems.

These and other fields started coalescing in the mid-1990s, developing common theories and experimental practices, to finally emerge as a single research discipline based on statistics, optimisation theory, and rigorous comparisons of performance. By the end of the 1990s, as the long funding winter was drawing to an end and expert systems were a distant memory, a new generation of AI researchers had a completely new set of tools and practices that allowed them to extract patterns and learn rules from large datasets and rigorously assess their predictive power. These researchers laid the foundations for the *lingua franca* of modern AI: a mathematical language made of different disciplines, and capable of naming the many ways in which machines can fail to generalise correctly. Their terminology and mathematical notations are still found in today's articles and patents, even in fields outside the sciences of computing.

For example: the entities or situations about which the intelligent agent is called to make a decision or a prediction are described by "features", annotated by "labels", split into "training set" and "test set", the result of learning is called a "hypothesis", and the archenemy is the insidious phenomenon called "overfitting" (described in Chapter 3). All those concepts, terms, and the relative theories behind them date back to the days of the great convergence that resulted in the modern field of machine learning.

This convergence came just in time, as the same years also saw the emergence of the World Wide Web, launched in 1994, which would completely transform the pursuit of AI. A marriage was inevitable.

SHOWDOWN IN SEATTLE

"Dearest AMABOT. If you only had a heart to absorb our hatred, thanks for nothing. You jury-rigged rust bucket, the gorgeous messiness of flesh and blood will prevail".

The atmosphere was tense at Amazon between the editorial and the automation teams in 1999, when this anonymous classified ad was published in the Seattle Weekly. It referred to Amabot, the software robot that was replacing human editors to populate the web pages that acted as shop windows for that new type of online shop. And the tension had been brewing for some years.

In the late 1990s, the online shop Amazon was just getting started, transitioning from a US-only operation to "the Earth's biggest bookstore". In this first incarnation (until around 1997 or 1998), Amazon had relied on a talented editorial team of some dozen people to generate high-quality reviews, inspired by the New York Review of Books, aimed at having a literary style that they called "the voice of Amazon". But this approach struggled to keep up with the rapid growth of the company; one former reviewer reports covering 15 non-fiction books per week. Soon the managers realised that they could not keep this up, as they rapidly expanded their catalogue to also include music, movies, and cover many more countries. All that was about to change when Amazon started to experiment with automated systems to produce book recommendations, eventually leading even to personalised recommendations.

The need for automation was real, as Amazon's founder Jeff Bezos was among the first to discover that scale was the key to the unexplored business of online shopping; but automated suggestions can be made in many ways, even a simple list of bestselling items can suffice. Amazon, however, soon started focusing on personalisation, pursuing Bezos' own "radical" idea of "one shop for each customer". This was a small but consequential shift, which moved emphasis from reviews to recommendations, and the company's automation and personalisation team was tasked with trying possible methods and algorithms.

A recommender agent is generally tasked with finding the items that are most likely to interest a given user. The first attempt at automating recommendations was based on a technology called Bookmatcher that involved users filling out a questionnaire about their reading preferences, which would be used to generate a customer profile and recommend to them what "similar customers" had already purchased.

Up to that point, the editorial team did not have much to worry about, but this was about to change. In 1998, researcher Gary Linden and his collaborators on the personalisation team developed a new algorithm; rather than finding similar users by profiling them, they thought they could just directly find similar products by analysing their database of past transactions. Essentially, similar products are those that tend to be purchased by the same people. This information can be computed in advance (and periodically updated) so as to be readily available to inform recommendations when a customer starts shopping. This approach to recommender systems was called item-based collaborative filtering, a tribute to a method invented by Dave Goldberg in the pre-web years to filter unwanted emails based on which users engaged with them: all users (sometimes implicitly) collaborated just by using the system. The method is discussed later in this chapter.

Rather than asking people what they thought or wanted, the system based its behaviour on what they—and millions of other people—actually did. When this method was tried, the results were clear: customers bought more books following personalised recommendations than following reviews written by people, and this made it possible to use algorithms to choose the content that would appear on the various pages. Taken together, these algorithms were called "Amabot", and the logo of the personalisation team indeed showed a stylized robot. The good performance of this "robot" is what sealed the fate of the small but brilliant team of human editors, who had been carefully developing "the voice of Amazon".

This is when the classified ad appeared in the Seattle Weekly, an act that we can consider as part of the same "culture war" we have seen unfold between Jelinek and the linguists at IBM ten years before, a conflict that would keep on returning in various forms in the following years (and we will meet again in later chapters).

As the Amazon editorial team struggled to keep up with their targets against the automated recommender software, the personalisation team had hung on the wall a reference to the folk ballad of John Henry, the legendary railroad worker who competed and won against the machine, only to die of exhaustion after that. The poster simply said, "People forget that John Henry dies at the end".

By the end of 1999, the editors were phased out while sales kept on growing, driven by an increasing reliance on personalisation and machine learning. The new Amazon algorithm was able to guess which products a customer was likely to buy; in the language of autonomous agents, Amabot could predict the consequences of its choices enough to act effectively towards its goal, and its environment was formed, or at least inhabited, by human customers.

The behaviour of the agent was not based on any explicit rule or understanding of customers or book content; rather it exploited statistical patterns discovered in the database of past transactions. Its "behaviour table" was neither obtained from exhaustively listing the correct action for each possible situation, nor from reasoning with explicit rules; it explored a space in between those two extremes, which enabled it to make good effective decisions in new situations based on previous experience. This agent pursued its aim of increasing sales, even though it did it in a different way than we would have; it was able to act rationally in a domain for which there was no theory.

The key intuition behind this method, that there are reliable patterns in human behaviour that can be learnt and exploited, is today behind many automated systems. On the one hand, this removes the need to ask questions of the users; on the other hand, it introduces the need to observe them. Today we receive personalised recommendations of videos, music, and news in this way, a method much closer to Jelinek's ideas than to Feigenbaum's. The recommender system is now the most common form of intelligent agent that we encounter online.

PARADIGM SHIFT

The events of the late 1990s at Amazon demonstrated how the World Wide Web could be a profitable environment for intelligent software agents to interact with and provided a viable model for other businesses to exploit that interaction. All they needed to do was to collect information about user behaviour and apply algorithms such as those of Amazon to exploit statistical patterns in that data, turning them into decisions. The infrastructure of the web would allow the agent both to sense its environment and to perform its actions, which in this case are recommendations; this resulted in goal-driven behaviour, capable of effective action in new situations, which one may call "intelligent". The success of this idea did not create just a new business model, but also a new scientific paradigm.

The history of science and technology is shaped by such success stories, which philosopher Thomas Kuhn called "paradigms". While investigating the historical "trajectory" of various scientific disciplines, Kuhn noticed that it is not one of smooth and constant progress, but rather one of sudden accelerations and swerves. He identified two different "modes" in these trajectories: the first he would name "normal science" and the second "paradigm shifts".

His key idea was that a "scientific paradigm" is formed by more than just a set of explicit beliefs held by practitioners of science about their domain: it also includes unspoken beliefs about what is the goal of their quest and the proper way to pursue it, what forms a valid solution and what forms a legitimate problem, and so on. These are often passed down to students in the form of examples, or important stories that act as models for imitation. It is these examples that are called paradigms, and they characterise an entire way to work.

In the history of physics, we can see various paradigm shifts, for example, in the transition from Newtonian to quantum mechanics, which comes with new expectations of what success should look like; the first is exemplified by the solution of problems such as a pendulum, the second by models of the hydrogen atom, where probabilistic predictions are all that one can aim for. The effect of a powerful success story of this type is to act as a model for future work. During periods of "normal science", researchers derive and polish results within the current paradigm, but very occasionally something happens and the paradigm itself shifts, and so do the language and the goals of that entire scientific community. This is what happened to AI at the turn of the century.

THE NEW AI IN THREE MOVES

As the dot-com bubble burst in March 2000, one of the great survivors was Google, which was emerging as the default in web search and had incorporated at its core the new paradigm, which would become known as "data-driven AI". Google would go on to become the dominant company in AI and set the research agenda for other companies and universities alike by identifying and solving a series of technical problems that were both profitable and just within the reach of this new technology.

Among the many innovations introduced by Google into their products over the years were machine translation across dozens of languages, query auto-completion and auto-correction, voice search, content-based image search, location-aware search, and—importantly—highly effective personalised advertisements, which went beyond what started by Amazon and which paid for much of the research.

By 2009, a brand-new culture had coalesced around Google and similar companies, and in that year a group of senior Google researchers published a paper that has since become a manifesto of that mindset. Named after a classic 1960 article by Eugene

Wigner about the power of mathematics, it was called "The Unreasonable Effectiveness of Data".

It celebrated the power of data in informing intelligent behaviour and codified what had become the common practices of the field into what can be seen as the recipe for a new AI. To make the point that it should be data, not models or rules, to inform the behaviour of intelligent agents, some of its most memorable statements include, "*Simple models and a lot of data trump more elaborate models based on less data*", and even the sentence, "*(Perhaps . . .) we should stop acting as if our goal is to author extremely elegant theories, and instead embrace complexity and make use of the best ally we have: the unreasonable effectiveness of data*".

More than twenty years had passed since the heyday of expert systems, and it did feel that way. The entire approach was conceived to operate in domains where there is no theory, replacing it with machine learning. The article does not directly mention Frederick Jelinek, but his presence is felt everywhere, for example, when it says, "*The biggest successes in natural-language-related machine learning have been statistical speech recognition and statistical machine translation*" and then proceeds to explain how this was entirely due to their effective use of data.

Replacing theoretical models with patterns discovered in data was *the first shortcut* taken along the way towards intelligent machinery, the same discovered both by Jelinek and by Amazon. One obvious problem of replacing theories with data, of course, is that of finding the necessary data, which in itself might prove to be as expensive and difficult as making the theories.

The authors of the article have an answer to that: use data already existent "in the wild", that is, data generated as a by-product of some other process. They explain that "(. . .) *a large training set of the input-output behaviour that we seek to automate is available to us in the wild. (. . .) The first lesson of web-scale learning is to use available data rather than hoping for annotated data which is not available"*.

This second shortcut, the same used by Amazon when it repurposed the sales dataset to discover which books were similar, takes AI very close to a free ride, replacing both theoretical models and expensive data at once.

One further step that is sometimes needed is that of annotating data with user feedback; how can the agent know what the user wants? Rather than asking them to fill out questionnaires (as was attempted by Bookmatcher), it became normal to just observe what they do, and infer their intentions and preferences from that. For example, in information retrieval tasks, the user might be presented with a series of news articles or videos to choose from, and their choice might be recorded as an indication of their preferences.

This third shortcut was also proposed at various times in the same period, for example, a 1996 article says, *"we make a design decision not to require users to give explicit feedback on which hits were good and which were bad (. . .) instead we simply record which hits people follow, (. . .) because the user gets to see a detailed abstract of each hit, we believe that the hits clicked by each user are highly likely to be relevant (. . .)"*. The expression "implicit feedback" dates back at least to 1992, when Dave Goldberg used it in his pioneering work on collaborative filtering of emails: after explaining that different readers can indirectly collaborate to filter spam *"by recording their reactions to documents that they read, such as the fact that a document was particularly interesting or particularly uninteresting"*, he adds, *"Implicit feedback from users, such as the fact that some user sent a reply to a document, can also be utilised"*. Ideas sometimes come before their time and have to wait for the rest of the world to catch up.

This way of designing intelligent behaviour is one of those cases where examples speak louder than words: imagine having to program a computer to recognise a funny joke or an unwanted email.

While a rigorous rule is unlikely to ever exist, providing examples can be easy. Of course, users do not respond very well to direct requests to provide information (see how we deal with cookie-preference messages), so it made sense to just learn by observing their behaviour. This practice amounts to using an observable "proxy" to stand in for the real signal that we intend to use, which is not directly observable—for example, user preferences—and has become widespread in the design of intelligent agents.

ON THE NEW AI

Where the "logical" AI of Ed Feigenbaum suffered from a chronic lack of "killer applications", the manifesto written by the three Google researchers spoke from a position of enormous strength, as it was reporting on the actual methodology that allowed this company to become the leading player in the AI market, as well as the whole domain of e-commerce.

At the same time, in 2009, Frederick Jelinek also received a lifetime achievement award from the Association for Computational Linguistics, and at that point nobody could miss that something deep had shifted: the definition itself of what we should consider as success in designing an intelligent system.

Twenty Five years before, when Ed Feigenbaum insisted that intelligent behaviour should emerge by performing logical inference from a declarative knowledge base, an example of good AI would have been a theorem proving algorithm, or one designed to perform medical diagnosis from first principles within a specialistic domain. In this tradition, progress in machine translation would have involved a breakthrough in our understanding of linguistics that could inform the reasoning of an inference engine of some sort.

This is how scientific paradigms change. By the time that manifesto was published, investors and students could easily contrast the lack of killer applications of "classical" AI with an abundance of working products coming from the data-driven camp: from language processing to the ubiquitous recommender systems, all success stories were powered by patterns learnt from data. A new generation of researchers had come of age in a world where data was readily available, and statistical machine learning had become the method of choice for producing intelligent behaviour, or at least one specific type of it. And the contents of university AI courses were changing as a consequence.

VAPNIK'S RULE AND THE NEW MINDSET

In 1990, the statistician Vladimir Vapnik left the Institute of Control Science in Moscow and moved to New Jersey, where he joined the prestigious ATT Labs. His work on the theory of machine learning reached back to the 1970s but it was just then becoming influential among computer scientists in the West, as they started to focus on a general theory for learning algorithms. Vapnik brought with him not only a deep mathematical theory of learning machines, which is still used today to understand the effect that changes to the machine could have on its performance, but also an attitude that resonated well with the technical developments of those days.

His approach ignored the question of how well an algorithm can identify the hidden mechanism that generated the data and its regularities, asking instead a different question: what predictive performance can we expect from a pattern that has been discovered in a set of training data? In other words: assuming that a regularity is found in a given set of data, when can we trust it to also be present in a future set of data? His theoretical models identify which factors should influence our confidence: the size of the dataset and the number of other possible patterns that have been implicitly tested and discarded during the analysis. Intuitively, the more hypotheses we try to fit to the data, the more likely it is that we hit one that just happens to do so by chance, and is therefore not useful for predictions. As esoteric as they seemed, Vapnik's formulas could be turned directly into very effective algorithms, and this is just what he did in the

American part of his remarkable career (Chapter 3 will discuss some of those ideas).

That approach abandoned the traditional task of identification of a hidden mechanism behind the data in favour of the simpler task of prediction of future observations. One piece of advice that he would often give to his students was, "*In solving a problem of interest, do not solve a more general problem as an intermediate step. Solve the problem you need, not the more general one*". The legendary statistician Leo Breiman once summarised this approach as "going right for the jugular". In other words, if you need to filter unwanted emails from an inbox, do not solve the general problem of language understanding; concentrate on the simpler problem of removing spam.

Once incorporated at the heart of a deep statistical theory of learning, this amounted to an epistemological, not just methodological, statement; all that matters is the resulting behaviour of the agent, so one should focus on that, be it a recommendation or a classification or a translation, without trying to solve some general problem such as "intelligence". If translations can be approximated without complete understanding of the phenomenon of human language, why would you need to make your life harder?

THE NEW RECIPE

These lessons accumulated during the 1990s, as the new generation of AI researchers faced the new challenges and opportunities presented by the web, and eventually shaped the whole culture surrounding the quest for intelligent machinery. The language of AI was no longer that of logical inference, but rather that of statistics and optimisation theory, and the main focus was on sourcing the necessary data. Machine learning became the central discipline of the entire field, the training data became the most precious resource, and performance measure became its obsession. The goal was not the discovery of any truth, but to generate behaviour that is "probably approximately correct", and for that simple statistical patterns are often enough.

If we combine together the maxims and teachings from the various researchers that have shaped the paradigm of data-driven AI, we can obtain a list of suggestions that reads like a recipe. Here is how they would sound if they were all concatenated.

When solving a problem, do not solve a more general problem as an intermediate step (Vapnik); fire a linguist (Jelinek); follow the data (Halevy et al). More data is more important than better algorithms (Eric Brill, cited by Jelinek); simple models and a lot of data trump more elaborate models based on less data. (Halevy et al). Memorization is a good policy if you have a lot of training data. (Halevy et al). Use data that is available in the wild rather than hoping for annotated data which is not available (Halevy et al). Do not require users to give explicit feedback, instead simply record which hits people follow (Boyan et al). Implicit feedback from users, such as the fact that some user sent a reply to a document, can also be utilised (Goldberg).

There is one more maxim that should be added to the above recipe: recent applications that require statistical modelling of natural language can only work once their training data exceeds one billion words, due to the large number of parameters that need to be tuned, a situation that some have summarised as "life begins at 1 billion examples".

Today, Amazon's recommendations are based on hundreds of millions of customers, those of YouTube on two billion users, and the most advanced language model of today—GPT-3—has about 175 billion parameters, which had to be learnt from analysing about 45 terabytes of text data obtained from a variety of sources. These models occupy an intermediate space between abstraction and enumeration, challenging what we traditionally meant by "understanding". But they are the only way we have found to operate in domains for which there is no theory, such as the prediction of human behaviour.

Vapnik and Jelinek had stumbled upon the same principles, which boiled down to this: rather than understanding the system under investigation, one can be satisfied with predicting what it will do. Predicting the next word in a text is much easier than understanding a sentence, and in many cases this is all that is needed. The same goes for labelling an email as spam or recommending a book to buy. Some called this "the end of theory", and its implications for the rest of science are not yet clear.

In discussing his take on learning from data, a process that he called Empirical Inference, Vapnik contrasted it with a famous quote by Albert Einstein: *"I'm not interested in this or that phenomenon (. . .). I want to know God's thoughts; the rest are just details"*. To this, Vapnik countered that the central question of Empirical Inference is "How to act well without understanding God's thoughts?"

This seems to be the ultimate shortcut that subsumes all those that we have listed above; and since biological evolution only sees and selects phenotype behaviour, should we not suspect that biological intelligence has been shaped more by its capability to help us to "act well" than any presumed capability to "understand God's thoughts"?

3

FINDING ORDER IN THE WORLD

Detecting regularities in the environment is a necessary step for an agent to anticipate the consequences of its own actions, so that a regular environment is a prerequisite for intelligent behaviour. There are various important limitations to any methods for pattern recognition, and they pose limits to what we can expect of intelligent agents too.

INTELLIGENCE, PATTERN, AND AN ORDERED WORLD

The sea slug *Aplysia* has a reflex to withdraw its delicate gills when touched, just in case this contact might signal that some predator is trying to nibble on them. But if subjected to a repeated series of similar touches, it slowly stops that reflex, a mechanism known as habituation, which is one of the simplest biological models of

DOI: 10.1201/9781003335818-3

learning. The utility of this mechanism implies the assumption that those repeated events do not signal any real danger to its gills.

Every environment presents different trade-offs between the costs and benefits of taking a risk, so *Aplysia* need to behave differently in each case. Biologists would have no further questions, seeing the evolutionary benefits of this simple type of learning; after all, if that contact has not led to harm for a while, it is probably not dangerous.

But this is where philosophers start becoming restless; is the "belief" of the slug, based on experience, ever justified? Scottish thinker David Hume in 1739 identified this as a key unsolved question and called it "the problem of induction". His conclusion was that there is no logical justification for the slug to base its expectations on experience, unless further assumptions are made, which in turn would be unjustified.

Philosophers have debated the question ever since, and in 1912, Bertrand Russell put it this way: "*Domestic animals expect food when they see the person who feeds them. We know that all these rather crude expectations of uniformity are liable to be misleading. The man who has fed the chicken every day throughout its life at last wrings its neck instead, showing that more refined views as to the uniformity of nature would have been useful to the chicken*".

Philosophers are known to ask awkward questions and take things to extremes, and this debate was about more than just animals; they were debating our belief in the natural laws that we discover from experience. Our belief in gravity is no different than the *Aplysia*'s belief that those contacts are innocuous, both being based on a finite amount of observations.

Higher animals are capable of learning much more abstract associations; for example, pigeons can be taught that pecking on the right images in the right order can lead to a reward. A group of pigeons, in New Zealand in 2011, were trained to order images based on the number of elements they contained: a set of two red ovals should come before a set of three blue circles, and so on. Once this skill was mastered, they were presented with new images which used set-sizes never seen before, containing up to nine elements; the pigeons were able to correctly order those too, showing that they could grasp the concept of number. While they were able to grasp a concept that would clearly be lost on an *Aplysia*, their justification for expecting rewards after taking a certain action was still based on the same leap of faith as the slug.

These are the two questions we face today when creating intelligent machines: as we expect them to go beyond memorisation and generalise to new situations, how do they know when they can trust a pattern that they have discovered in their past observations? And how do they know that they are not missing some useful pattern, just because they have not noticed it?

For a philosopher, this question can be turned around: why should the environment be so simple and stable that we can both grasp its regularities and trust they will be there tomorrow? It is only in those environments that an agent can anticipate the future and therefore behave rationally.

Machine learning, the science of building machines that can transform past observations into knowledge and predictions, faces the same questions posed by philosophers of science centuries ago. How can we trust the knowledge produced by our machines? Should we, as humans, expect to understand it? Could we one day find ourselves in the position of the *Aplysia,* unable to understand what is perfectly clear to a pigeon?

There are different, and equivalent, ways to understand what we mean by a "pattern" in our data. Consider the object in Figure 3.1, which we will call Pascal's triangle, but was known in Persia and in China for centuries before Pascal was born.

Every entry of the triangle can be reconstructed from the other ones, specifically by taking the sum of the two entries that are directly above it. Once you notice this pattern, you can fill in any missing values, detect any errors, or generate one more row. In fact, you can generate as many new rows as you want, since we are showing just a part of an infinite object.

Importantly, we can compress this entire structure into a single mathematical formula based on the property that we have

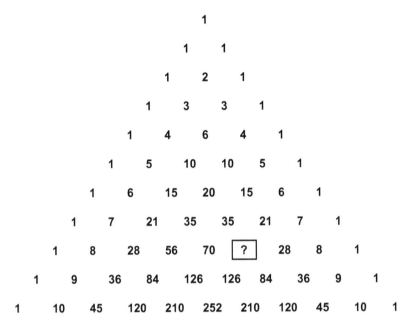

Figure 3.1 For the readers who are familiar with the factorial notation, the entry in position (n,k) can be calculated as: $n!/k!(n-k)!$, where n=0,1,2 . . . indicates the row, and k=0,1,2 . . . indicates the column. This formula is a finite description of an infinite object.

just described, or in a few lines of computer programme. This is the key to understanding patterns; rather than listing all the values in the table, we can describe them.

We will think of a pattern as anything in the data that can be described, and we will be particularly interested in the case when an infinite object allows for a finite description.

Once you can describe a set of data, you can use the same formula or computer code to fill in any blanks, extend the set of data that you already have, or even spot any mistakes in that data; they are the few points that violate the simple pattern that you have just discovered. When such a relation is discovered in the data, we normally say that we have "learnt it" from the data. (Note that although we make an example of an exact pattern, similar discussions are possible also for approximate patterns).

But here is a catch: not all sets of data can be learnt; for example, a random permutation of the entries in the above figure would not be learnable: you would not be able to reconstruct missing entries, predict its next row, or describe it in a few lines of computer code. You would have to list or memorise every single entry in every single position.

Mathematicians can show that environments that behave like Pascal's triangle are a tiny fraction of all conceivable environments, and all the rest cannot be described in abstract terms. Learning and intelligent behaviour are possible in a small set of worlds, leading some philosophers to wonder why we happen to live in one of them. Why can *Aplysia* benefit from habituation, and physicists describe the past and predict the future using laws and theories?

One possible answer is that *Aplysia* and physicists would not be able to exist in any other environment; so long as there is an intelligent entity looking at it, its world must be a predictable one. This tautology is known as the anthropic principle in the philosophy of science.

The life of a scientist can be hard; if a pattern is not detected, does it mean that the data does not contain any? And if it is detected, was it a coincidence, or can we trust that it will still be there tomorrow? In both cases, the answer is negative; failing to find a relation cannot prove that the data is random, and finding it does not ensure that it will be there in the future. In both cases, we need to take a leap of faith, which is what usually upsets philosophers.

Could the scientist be helped by a machine? *Machine learning* is the technology of detecting patterns in data, extracting relations that we can trust, while avoiding being fooled by coincidences, in order to make reliable predictions. As we have seen in Chapter 2, theoreticians of machine learning today think that the machine-learning game should just be about prediction, not the identification of any deeper truth. They also accept other limitations, for example that they might be unable to interpret whichever patterns their machines have found.

LIMITATIONS TO KNOWLEDGE

Many things can go wrong when a person or a machine attempts to find order in a set of observations. Some of them have implications for the way we should relate to the intelligent agents we are building and can be illustrated by considering simple questions. These questions are: 1) whether there can be a general method to detect any pattern in any dataset; 2) how we can trust that a relation found in the data is not just the result of a coincidence; and 3) what we should expect to learn from inspecting the relations discovered by the machines.

No Free Lunch

Many algorithms are guaranteed to always find the intended solution, for example, the shortest path between two points on a street map or the product of two integer numbers, regardless of the specific input. So it may be natural to expect a single universal algorithm to detect any pattern in data, but the general consensus is that this cannot exist: for any algorithm that can learn patterns in data, there will always be some data that appears random to it, while being learnable by another algorithm. This situation resembles the moment in physics when we finally accepted that we cannot generate perpetual motion: while disappointing, this was also important to make progress. If it existed, such a procedure would solve a number of problems in computer science involving data compression and randomness checking, but at least we can focus our efforts on developing more specialised algorithms: as soon as assumptions are made about the source of the data or the type of pattern being sought, often we find powerful methods to find them. For example, we have very efficient methods to detect periodic structures in time series, or linear relations between variables, and many other useful patterns.

Coincidences

Once a relation in a set of data has been found, we need to know if we can trust it to still be there in another set of data from the same source. What if that was just a coincidence? This is a real possibility when the dataset is small, or we are allowed to cherry-pick our data or the relations that we are interested in, but it seems to be surprisingly difficult for some people to appreciate, as we can see in this old story.

The August 21st, 1964, issue of *Time* magazine featured a short article entitled "A compendium of curious coincidences" that rapidly attracted significant attention. President Kennedy had died just nine months earlier, and the article reported on a story that had already been circulating around Washington, DC for some time, about a strange series of coincidences that related him to his predecessor Abraham Lincoln.

"Lincoln was elected in 1860, Kennedy in 1960. Both were deeply involved in the civil rights struggle. The names of each contain seven letters. The wife of each President lost a son when she was First Lady. Both Presidents were shot on a Friday. Both were shot in the head, from behind, and in the presence of their wives. Both presidential assassins were shot to death before they could be brought to trial. The names John Wilkes Booth and Lee Harvey Oswald each contain 15 letters. Lincoln and Kennedy were succeeded by Southerners named Johnson. Tennessee's Andrew Johnson, who followed Lincoln, was born in 1808; Texan Lyndon Johnson was born in 1908".

Since then the list has kept circulating widely, gradually growing, even if not all facts included in it were entirely correct. Some were simply wrong, others were stretched to the point of becoming absurd. But the story of the eerie connections kept on travelling, and the list kept growing.

What some people find difficult is to answer two questions: even if the information was entirely accurate, would those connections be surprising? Do they signal some extraordinary connection between the two presidents that is deserving of an explanation, or that can be used to guess further information about them?

Statisticians can settle this by asking two further questions: for how many other pairs of people could we have generated a similar list? And how many other facts about those two presidents could we have added to that list if they had been true?

The above compendium of curious coincidences is the result of cherry-picking, by which all information that shows a connection has been kept, and anything else has been removed from the list. There is no mention of their city of birth, shoe number, and so on. Patterns that can be explained as the result of chance are considered "non-significant" and should not be used for predictions or as part of scientific theories. If an explanation can always be found, because we can reject the data that does not fit our theory or we accept extraordinarily convoluted explanations, then the result of the pattern detection process is not informative. We call those patterns "spurious", the result of the much-studied phenomenon of "overfitting", which occurs when the hypothesis attempts to explain accidental correlations in the data.

Statistics, machine learning, and the scientific method have all studied this risk and have come up with rigorous ways to avoid it, which are today built into the software we use to train our intelligent machines. However, if the human operator does not understand these principles, they can still end up with machines that dream-up non-significant relations in data. This risk is very present when we ask our machines to generate massive and unreadable models to make predictions about domains that we do not understand.

Alien Constructs

One standard way to compress a large table of observations, or a behaviour-table containing stimulus-response pairs, is to introduce abstract terms that describe groups of objects and then describe the data using those terms. Consider a long list of animals and the food that they can eat:

ANIMAL	FOODSTUFF
tortoise	lettuce
cat	fish
rabbit	carrot
. . .	. . .

For a sufficiently large table, this could be compressed by creating the concepts of *herbivore* and *carnivore*, in the first column, and the concepts of *vegetables* and *meat* on the other. These terms do not describe an object but a category of objects and are the foundation of a theoretical description of the world, enabling us to state simple and general rules such as "herbivores eat vegetables".

Learning machines create constructs too; for example, a video recommender system will be able to divide both users and videos into classes, and then use that terminology to express the relations it observes in user behaviour. While being useful to generalise, those constructs do not need to be meaningful to us.

For example, in one of his most famous stories, Jorge Luis Borges tells of a fictional world where animals are classified as follows:

1. Animals belonging to the emperor
2. Embalmed ones
3. Trained ones
4. Piglets
5. Mermaids
6. Fabulous animals
7. Stray dogs
8. Those included in the present classification
9. Those that (shake) as if they were mad
10. Innumerable ones
11. Those drawn with a thin camel hair brush

12. Others
13. Those that have just broken a vase
14. Those that seen from afar look like flies

These classes would probably be useless to describe the table of eating habits or to summarise phylogenetic or ecological information, but might well be useful to describe some archaic legal system where the animals of the emperor or stray ones have special rights. This example may look artificial, but it is not too distant from the "segmentation" methods described in later chapters, where a population of individuals is analysed within the context of direct marketing, individual risk assessment, or content recommendation systems.

There is no "objective" way to divide and describe the world, and Borges himself notes in the same essay, "there is no classification of the universe that is not arbitrary". There is also no reason to expect that two artificial agents observing the same world would come to the same theoretical abstractions, even when they can make the same predictions. Understanding their reasons in our language might well be impossible.

PATTERNS AND COGNITIVE BIASES

There is another reason why it is difficult for us to handle cases where spurious patterns have been found: people have an instinctive drive to seek patterns in the world. While this is one of our most valuable traits, it can also lead to failures too, particularly when we are confronted with a lack of pattern—that is, randomness. Both people and machines can be fooled by coincidences, or by our own cherry-picking, and while we may think that our theoretical constructs reflect some hidden reality, they are often just useful shorthands to simplify our descriptions of the world. Just as there are no general-purpose learning algorithms, so humans are not general-purpose learning systems either.

The human tendency to impose interpretations on random inputs is called *pareidolia*, and in some cases this can become pathological.

People see faces in rocks, the moon, or even toast. A more abstract version of this tendency to perceive meaningful connections between unrelated things is called *apophenia*. It is possible to see causal connections between completely independent events. In pre-scientific times, it was common to see higher meaning in events that can readily and trivially be explained as the effect of chance, and even today this tendency is still present when we explain reality in terms of complicated conspiracies or machinations.

As we look at the night sky, we cannot help noticing the triplet of equidistant stars that form Orion's Belt. It seems that all cultures had a name for it, and possibly humans were aware of it from their very origins, simply because the same mutation that made it impossible for us to miss that pattern might just have been what made us humans. But there is no structure in the positions of stars: we can memorise them, but could not reconstruct a missing part of the sky from the rest. Are the stick figures that we see in the sky hallucinations?

We do the same thing for causality; there is a deep difference between the time series of day length and that of lottery numbers, the first being easily compressed and predicted, the second being entirely random. Yet many players believe that long overdue numbers are more likely to appear next, or numbers that have some connection with events in their lives.

WHAT MACHINES CAN DO

Researchers in machine learning have largely dropped any ambition of "identification" of the hidden mechanisms behind the data in favour of the weaker goal of making useful predictions. In this way they have moved one step closer to modern philosophy, but also a step farther from us being able to relate to our machines. Not only should we expect machines to discover patterns that we cannot comprehend, and to make predictions better than us in certain domains, this is already a reality today.

Consider the case of GPT-3, a language model created by OpenAI in 2020. Language models are tools to compute the probability of a sentence in natural language, and used mostly for text prediction tasks, which include completing sentences, filling in blanks, and suggesting changes. Statistical language models are commonly used as part of many systems, from text correction to autocompletion to question answering and even simulating dialogue. The vast model was trained on 45 terabytes of raw text gathered from a variety of online sources, which would require over 600 years to be read by the fastest human reader (Mr Howard Berg, who is listed in the Guinness World Record for having been able to read 25,000 words per minute). The model itself contains billions of parameters, which is way more than we can comprehend in our working memory.

These models are the direct descendants of the language models introduced by Jelinek in the 1970s, and one of these, called LaMDA and created by Google, ended up in the newspapers in 2022 because an engineer became convinced that it has become "sentient" after a long "conversation". LaMDA has been trained on four billion documents to simulate dialogue and contains billions of parameters.

I do not know of studies which attempted to understand the world as it is represented within that model, but even if the model was small enough to be meaningfully inspected, there is no reason to expect that its internal abstractions would match those that we use in our life.

Not only do GPT-3 and LaMDA have superhuman size and experience, they may also have "alien" representations of the same world, so that they would perform the same task in very different ways from us, and from each other. As machine learning pioneer Vladimir Vapnik pointed out, if the goal of the game is prediction, there is no need for the machine to solve a harder problem as an intermediate step.

UNDERSTANDING OUR CREATIONS

Goal-driven behaviour requires the capability to anticipate the consequences of one's actions, and a reliable way to achieve this is by adaptation, or learning. This is what the *Aplysia* does when it habituates to a stimulus, whether it has reasons or not to expect its environment to be regular.

This is also the main way in which we create intelligent agents: by training them on vast amounts of data so that they can detect useful patterns that they can use to inform their decisions. From video recommendation agents to spam filters, they all work in the same way. Can we trust that they will always work? It will depend on the nature of the assumptions that we implicitly make when we create them.

Fortunately for the AI industry, there seems to be a surprising amount of order in human behaviour, even if we do not have a theoretical model of it and exceptions abound. Models that are based on a mix of listing observations and measuring simple statistics seem to go a long way towards predicting our language and actions, a fact that is perhaps one of the most overlooked scientific findings of the past decades. But all these models require super-human amounts of experience, before they start producing useful behaviour; as some researchers say, "Life begins at one billion examples".

The revolution started by Jelinek is implicitly founded on this observation: large amounts of data and non-theoretical models of the world can generate useful behaviour, even though they may teach us nothing about the actual phenomenon. So we may not have a way to interpret the decisions of our machines, which would have been handy, in order to check that they do not go astray.

We should keep this in mind, as we consider ways to teach machines our values and norms; it might be like trying to explain them to a slug, or a clever pigeon. As Ludwig Wittgenstein once put it, "*If a lion could speak, we would not understand it*".

4

LADY LOVELACE WAS WRONG

*Machines today can do things that we cannot do or even under-
stand, thanks to their ability to learn them autonomously from
experience. This makes them suitable to operate in domains for
which we have no theory, and may be the main advantage they
have over traditional methods.*

THE LADY AND THE MACHINE: FROM NOTE G TO MOVE 37

London, 1843

In 1843, Lady Ada Lovelace had just finished translating into
English a technical treatise by Luigi Menabrea, which was based
on the lessons of her friend Charles Babbage, and decided to
add some ideas of her own. Even before mentioning the topic
of her work, this story may sound implausible due to the excep-
tional individuals involved.

DOI: 10.1201/9781003335818-4

Babbage was a prototypical Victorian inventor and polymath, famous during his lifetime, who constantly focused on multiple projects from railroads to insurance companies. He is now mostly known for his mechanical calculators, a pursuit that led him to conceive increasingly advanced devices until he invented "the Analytical Engine", the first general-purpose computer. It was just this machine, imagined a century before Alan Turing's creations, that was the topic of Lady Lovelace's publication. But Ada Lovelace was not just a close collaborator of Babbage's; she was also a socialite, the daughter of romantic poet Lord Byron, and an accomplished mathematician.

The book she translated from French in 1843 had been written by Luigi Menabrea, after attending a series of lectures held by Babbage in 1840 in Turin, where he described the Analytical Engine in detail. Years later, Menabrea would go on to become a Prime Minister of unified Italy, but at that time he was still a young officer of the Kingdom of Piedmont, fascinated by calculating machines, who happened to write the only account of Babbage's invention.

As remarkable as these connections are, they are not the most important part of the story: that is in a series of long notes that Ada added to the work as a way to include her own ideas on the nature and the potential of computing machines. Because of these extensive additions, we remember her today as the first computer programmer, even though the machine was never fully built.

Of these notes, it is Note G that has attracted much attention. It says, among other things, *"The Analytical Engine has no pretensions whatever to originate anything. It can do whatever we know how to order it to perform"*.

This reassuring thought would be repeated over and over again in the decades that followed the invention of electronic computers, to soothe our recurring apprehensions about the potential risks of that new technology by insisting that computers can only do what humans have programmed them to do. The only problem is that Ada Lovelace was wrong.

Seoul, 2016

DeepMind is a subsidiary company of Google, located just over a mile from Ada Lovelace's London house in St James Square, and whose stated mission is to develop a general form of AI. In 2016, the company combined two different techniques from machine learning to create a powerful new algorithm that could teach itself to play Go by simply playing millions of games with itself, reaching incredibly high levels of performance.

In doing so, they embarked on a trajectory that would bring them to collide with Ada Lovelace's maxim on the fundamental limits of computers, and the expectations of many experts. That algorithm was named AlphaGo.

At first, the game of Go does not appear too different from other board games, with two players taking turns to position black and white pieces on a 19×19 board, with the aim of capturing the pieces of the opponent. But the number of possible moves and responses at each iteration is so large that brute-force plans do not work as they do for chess; success does depend on knowing how to evaluate the quality of a position, what is called the "evaluation function". Since there is no general theory of how human champions do this, the machine is expected to learn it from experience by analysing tens of millions of recorded matches and then playing many millions more matches against itself, each time adapting its evaluation function.

By the spring of 2016, the algorithm had already defeated its programmers, as well as the best player in Europe (Fen-Hui), and it was time to take on the heavy weights. Korean Lee Sedol was the most prominent Go player in the world, and DeepMind wanted its software to play with him. The five matches took place in Seoul over one week under significant media attention.

It might not be surprising to find that the machine did beat the best human player; after all, this has happened with chess (when world champion Gary Kasparov was defeated by Deep-Blue in 1999) and all other main board games.

The point is what happened on move 37 of the second match.

This is when AlphaGo made a decision that all programmers considered a mistake, and so did its opponent. Nobody could interpret it, which means that nobody could see its purpose, or what function it served in the quest to gradually constrain the opponent and ultimately capture its pieces. But it turned out that the move was made with purpose, as it later became the very foundation of the machine's final attack that resulted in victory. Not only could the machine do things that none of its programmers could do. The machine could make moves that nobody could even interpret.

The knowledge of AlphaGo did not come from its makers, but from observing 30 million recorded matches and playing 50 million more games with itself, an amount of experience that would require more than a lifetime to a human player. And this made it behave in ways that its makers could not understand, or, in the words of Lady Lovelace, it let it do something that we do not "*know how to order it to perform*".

Where did this leave us and our illusions of inherent superiority?

MACHINES THAT LEARN

What Lady Lovelace—and everyone else at her time—could not have imagined was that in the years between her Note G and that fateful match in Seoul, engineers would develop the method of "machine learning", the technology of machines improving their performance with experience. It was through machine learning that AlphaGo had become capable of a purposeful behaviour that was both superior to that of its makers, and not understandable by them.

Let us take a minute to consider how many different tasks we describe as "learning" in everyday parlance. We can learn a phone number, a poem, a new language, how to ride a bicycle, how to recognise a type of poisonous mushroom, how to play chess and even Go. Despite having the same name, these tasks are not the same: some of them involve nothing more than memorization, while others require extrapolating from examples so as to generalise to new cases, such as when we learn to recognise a poisonous mushroom. And these are just a few of the possible uses of that word.

We do not consider part of "machine learning" simple tasks of data memorization, which is a domain where machines are already vastly better than us, but we reserve that term for situations where the machine can acquire new skills, or improve old ones, without being explicitly programmed. It turns out that not only is this possible, this is the only way we can make machines that can recognise a poisonous mushroom, win at chess, ride a bicycle, or recommend a video. All intelligent agents in use today are based on some form of machine learning; this is not just what allows them to cope with changing environments, but also with domains which do not have a clear theoretical description.

From a big distance, most learning algorithms work in the same general way, which can be clarified with an example. An algorithm and a cooking recipe have a lot in common, both specifying a series of steps to be followed in order to turn an input into an output.

The cooking recipe will include a list of ingredients and a sequence of actions, and each of those will require further details, such as quantities, temperatures, or durations. For example, a simple recipe might say: mix 60 grams of water with 100 grams of flour and bake at 200 degrees for 10 minutes.

Any change to those specifications might yield a different result. Even if we keep the same ingredients and operations, we can tweak the four numeric values of the recipe, to experiment with the space of possibilities. An experienced cook will have tried this many times, reaching an optimal setting; still, they might want to experiment every time they work in a new kitchen or use a different brand of flour. Feedback will be needed, in terms of someone tasting the food, usually the cook themself.

In mathematical language, these quantities of the recipe that one can adjust are called "parameters", and it is common to borrow language from music to describe these adjustments as "tuning the parameters", just like you may tune a piano. Feedback here comes from the refined ear of the piano tuner.

This "tuning" is a standard way for machines to learn (that is, to change their behaviour based on experience), and can be applied to the numeric parameters that control the predictions of agents used to recommend movies, words, or chess moves. Feedback from the environment is used as a guide towards the optimal values, which are those that produce the most effective actions.

Rather than listing the behaviour of the agent by an explicit table of stimulus-response pairs, it is normal to specify those pairs implicitly, in terms of a calculation that depends on tunable parameters, and then let the machine figure out the details from its interaction with the environment. The result will be its very own behaviour, possibly an original one that its maker might not have imagined.

Of course, learning is not just done by blind search, and a very deep mathematical theory exists for how to optimally adapt the settings of an algorithm in order to enable it to best pursue the goals we give it. This is what happened with AlphaGo.

So, was Lady Lovelace right or wrong when she said, "*The Analytical Engine has no pretensions whatever to originate anything. It can do whatever we know how to order it to perform*"? Is finding the best version of a complex recipe equivalent to "*originating*" something or is that just doing "*whatever we know how to order*"?

This may be a question for the philosophers, but I think that when you have many millions of interacting parameters, whose effect on the resulting behaviour you cannot predict, then the machine is indeed learning something that is new to you, and I would call it a new recipe. And when this ends up solving a problem for which you have no theory, such as winning at Go or recommending a book, it is appropriate to talk about autonomous goal-driven behaviour and also of learning and intelligence.

THE IDEA OF A LEARNING MACHINE

AlphaGo was not the first programme to teach itself how to play a board game better than its maker, becoming famous while doing so. In 1956 American TV Audiences were introduced to Arthur Samuel and his checkers-playing program which had learnt to play from examples listed in a classic game book called *Lees' Guide to the Game of Draughts* and through self-play, to the point of beating Samuel himself at the game.

In 1962, Samuel (or more probably his employer, IBM) decided that it was time for the machine to face a stronger opponent, and they chose a blind player from Connecticut called Robert Nealey who was to become the state champion in 1966, but who was not particularly well known at the time of the match, despite being billed as one of the foremost players in the nation. Using a new IBM 704 computer, the program defeated Nealey and the media loved it.

The article in which Samuel described the technical details of the algorithm was published in 1959 and had a remarkable—if delayed—influence on modern AI; not only did it introduce various methods that have since become standard, but it even first introduced a fortunate new definition for its field of study. That article was entitled *"Some Studies in Machine Learning Using the Game of Checkers"*.

In it, Samuel observed that there was a whole class of computational problems where we know exactly what we want, but have no mathematical rule to obtain it, and therefore require extensive manual tuning when implemented in machines. About automating their solution, he wrote *"It is necessary to specify methods of problem solution in minute and exact detail, a time-consuming and costly procedure. Programming computers to learn from experience should eventually eliminate the need for much of this detailed programming effort"*.

Board games were offered as an example. In this case, the machine will look ahead, from any given game configuration, to see what the possible outcomes of a move are, but eventually it will need a way to decide which "future" it prefers. To continue with our cooking recipe analogy, the "tunable parameters" of a chess or checkers playing agent would be the "weight" it will give to various features of a board configuration: how desirable is it to control the centre, or to have more pieces, and so on. Each choice of those quantities would result in a different behaviour. Samuel's key contribution was to show practical ways to adjust those parameters in a way to improve the chances of winning.

In July 1959, the New York Times reported Arthur Samuel as saying that *"machine learning procedures may greatly diminish the amount of information that must be given to a machine before it can work on a problem"* before adding, *"such an achievement may be twenty to fifty years off"*.

Today, recommender systems observe their users to learn how much each feature of a video contributes to its "clickability": a given genre, language, duration, or perhaps a word in the title. In the absence of a theory of user preferences, machine learning allows them to automatically look for the right proportions of each ingredient that increase the probability of user engagement.

Samuel concluded that such a system can be made to improve its game *"given only the rules of the game, a sense of direction, and a redundant and incomplete list of parameters which are thought to have something to do with the game, but whose correct signs and relative weights are unknown and unspecified"*.

The 1959 article summarised the lesson of this experiment as, *"a computer can be programmed so that it will learn to play a better game of checkers than can be played by the person who wrote the program"*.

Today we are still digesting the consequences of this conclusion. Samuel passed away in 1990 before seeing his legacy being picked up by AlphaGo. I have often wondered if the authors of AlphaGo remember that in Samuel's paper, the two computer programs facing each other during self-play were named Alpha and Beta.

SUPERHUMAN: ALPHAGO AND ITS DYNASTY

In the six decades that followed Samuel's experiments, the fields of computer science and AI were transformed beyond recognition; the first steadily followed Moore's law of exponential improvements in hardware, the second went through various hype cycles and dramatic changes of direction. Many new AI methods were discovered, developed, discarded, and sometimes rediscovered during these years.

The algorithm that defeated Lee Sedol at Go was run on hundreds of processors (1,202 CPUs and 176 GPUs) and had been trained on a database of recorded historical games, using around 30 million moves extracted from 160,000 different games, and then further trained by playing against itself 50 million times. Its training had taken weeks, but the same amount of experience would have required much longer than the lifetime of a human player.

There may be various ways for a machine to achieve superhuman performance at a given task, and many may feel a bit like cheating: one of them is to leverage more experience than any human can; another is to rely on larger memory and faster computation; yet another—not applicable to board games but to many situations—is to have superior senses, that is, access to more information than the human adversary.

AlphaGo was obviously superior in all possible dimensions to the program written by Arthur Samuel, including the hardware, the difficulty of the game, the way to look ahead, the number of parameters to be tuned, the number of training examples, and the board evaluation function.

It is unfortunate that all the knowledge distilled by AlphaGo during all its training will not be readable by human experts, as it is distributed in millions of numeric parameters which have no meaning to us. The way the machine carves its world to think about the next moves might well be similar to the example in Chapter 3, about "alien concepts" and the animal taxonomy of Jorge Luis Borges.

Later versions of the AlphaGo algorithm beat earlier versions and were not even trained by and deployed against human competitors. The latest one did away with the initial set of 30 million human games, learning entirely from scratch through self-play; it still defeated the original AlphaGo in 100% of test matches. The explanation of its performance provided by its programmers was as matter-of-fact as it was chilling; "it is no longer constrained by the limits of human knowledge", they said. Further versions, called AlphaZero and MuZero, taught themselves a series of board games, always achieving human or superhuman levels.

THE CREATURE OUTSMARTING ITS CREATOR

Over the years, Lady Lovelace's comment about computers being unable to "originate anything" was often repeated to allay our periodic anxieties about our increasingly powerful creatures. In the light of AlphaGo and its successors, should we worry?

There is no shortage of literary stories about creatures outwitting their creators, and there was even a philosophical debate in 1952 in The British Journal for the Philosophy of Science which included an article by cyberneticist Ross Ashby entitled, *"Can a Mechanical Chess-Player Outplay Its Designer?"*, which answered the question in the positive. It is natural to worry.

I do expect to be outwitted and outplayed in a number of important tasks, but what I do not find scientific is the possibility of a general-purpose or "universal" form of machine intelligence, which reminds me of those old illustrations of an evolutionary ladder leading from a tadpole to a human. The term "generalist" is more acceptable in this sense, as opposed to "specialist", since it still allows us to imagine several non-overlapping or partly-overlapping generalist agents.

Intelligence has inherently many dimensions, and different agents might be incomparable, unless we think we can compare the cognition of a crow with that of an octopus. So while I do expect machines to exhibit "superhuman performance" at many tasks very soon (from finding Waldo to driving a truck), I still do not find it useful to talk about "universal intelligence" as an absolute category, nor of a single general type of "superhuman" intelligence. We can still worry of course.

Superhuman performance in specific domains could derive simply from better senses and larger memories, but probably also from having access to a superhuman amount of experience, as we saw in AlphaGo and GPT-3. We will discuss in Chapter 8 an algorithm that can teach itself to beat human players at 57 different video games. We also already know of AI algorithms that can compete with human radiologists in the examination of chest X-rays, and one day the competition might well be over.

There will be areas where humans will hold their own. As French philosopher Claude Lévi-Strauss once noted, the important part of science is to pose the right questions, not to give the right answers. And this is what Lady Lovelace ultimately did for us in Note G: she addressed the question of what machines can or cannot do. Today's computers could easily translate Luigi Menabrea's book from French into English for her. But that is where they would stop.

5

UNINTENDED BEHAVIOUR

Can we trust that artificial intelligent agents, trained on data from the wild, will do what they are expected without causing collateral problems? The founder of Cybernetics, Norbert Wiener was concerned about machines taking harmful shortcuts and compared them to the talisman of an old horror story which granted wishes, except that it did so in a "literal minded" way. Today his concerns could become real: we use statistical algorithms for individual risk assessment, and base high-stake decisions on their output. Fortunately, so far, the damage has been contained, but how do we make sure that our machines do not violate fundamental social norms, by doing what they are asked to in a "literal minded" way?

THE MONKEY PAW

"I wish for two hundred pounds", said old Mr White, on the insistence of his family, although he did not really believe that the

DOI: 10.1201/9781003335818-5

old, mummified monkey paw had any magic powers. It had been difficult to come up with a single wish, let alone three, and it was his young son Herbert who thought first of the mortgage. "If you only cleared the house, you would be quite happy", he said; "well, wish for two hundred pounds, that will just do it". So Mr White, still sceptical, held up the talisman and spelled out this very specific request.

The next day he was no richer and still thinking about the old friend who had assured him of the powers of the paw, when a hesitant man knocked on his door. Mrs White saw him first, and immediately knew something was not right.

"Is anything the matter?" she asked when the man came in. "Has anything happened to Herbert? What is it?"

"He was caught in the machinery," said the visitor in a low voice.

"Caught in the machinery," repeated Mr White, dazed, "yes".

"I was to say that Maw and Meggins disclaim all responsibility", continued the visitor. "They admit no liability at all, but in consideration of your son's services, they wish to present you with a certain sum as compensation".

Mr White dropped his wife's hand, and rising to his feet, gazed with a look of horror at his visitor. His dry lips shaped the words, "How much?"

"Two hundred pounds" was the answer.

Norbert Wiener, the founder of cybernetics, was very fond of this story (abridged and simplified in the summary above) which was written in 1902 by William Wymark Jacobs, known as W.W. Jacobs. He also mentioned it in his prophetic 1950 book, *The Human Use of Human Beings*, to illustrate the possible risks of intelligent machines. In that book the "*monkey paw*" is used as a way to represent a device that pursues its given goal in a "*literal-minded*" way, indifferent to any harm that might result from its actions.

Seventy years later, we have reached the point where Wiener's concerns are justified, and we are fortunate that so far, the damage has been contained.

THE EXAMPLE OF INDIVIDUAL RISK ASSESSMENT

The shortcuts that enabled AI to rapidly flourish in the past 20 years are made possible by machine learning algorithms and large samples of human behaviour. Once trained on superhuman quantities of data, algorithms such as GPT-3 can complete sentences, predict purchases, answer questions, and assist with translation, all without an understanding of the subject matter. There is no theoretical reason why it should be possible to emulate or predict that behaviour using statistical means, but it is a remarkable experimental finding that psychology probably still needs to fully absorb. Similarly, with the right amount of data, a recommender agent can correctly guess which videos we will be likely to watch or which news we will want to share with our friends. Can that power apply to other aspects of life?

It could be that we are not as complex as we like to believe; maybe for a complex enough agent, we are not too challenging a "task-environment", and we never noticed this before because it was just impossible to gather large enough samples of behaviour. But even if that was true, and personalised predictions of future behaviour were possible, could they cross some of the cultural lines that we have come to value?

The possibility to predict the future behaviour of a person, even just statistically, is very tempting for a vast industry that existed for much longer than AI: the industry of individual risk assessment. This includes insurance, lending and credit, recruitment, admissions to education, employment, and even judicial (and psychiatric) situations. In all of these cases, an assessment is done to decide how an individual is likely to behave or perform in the future, so as to inform a present decision.

Traditionally, this has been based on actuarial methods, psychometric ones, or interviews at assessment centres. But there is currently a trend in devolving many such decisions to AI algorithms, due to their capability to make reasonable guesses about future behaviour.

Individual risk assessment is also one of the places that is most likely to provide us with a case of the monkey paw, because getting those predictions right is only part of the task; they also need to respect a large body of social norms that are very difficult to formalise, but nevertheless are requested by law and conscience. This is a place where context matters a lot; the very same decision can be acceptable, or not, based on the reasons for making it.

REGULATED DOMAINS AND PROTECTED CHARACTERISTICS

The reason why individual risk assessment is such a delicate area is because it is often used to inform decisions about access to opportunities, and often those are regulated from the point of view of equal treatment.

For example, in the UK, the Equality Act of 2010 requires that there is equal access to *employment, public services, healthcare, housing, education and qualification, transportation, and public authorities.* Every country has specific provisions to ensure equality of treatment for its citizens when it comes to those important areas of life. The same act in the UK states that "*The following characteristics are protected characteristics: age; disability; gender reassignment; marriage and civil partnership; pregnancy and maternity; race; religion or belief; sex; sexual orientation*".

In the European Union, the Charter of Fundamental Rights, which is legally binding since 2009 (Lisbon Treaty), has a section on equality: "*Any discrimination based on any ground such as sex, race, colour, ethnic or social origin, genetic features, language, religion or belief, political or any other opinion, membership of a national minority, property, birth, disability, age or sexual orientation shall be prohibited*". The USA has similar laws, though not yet unified into a single act.

This means, for example, that access to education cannot be affected by a person's religion or sexual orientation, or that employment decisions cannot be shaped by the race of an applicant. This also means that deployment of intelligent algorithms to make decisions in those domains carries the risk of violating some law. This is where an AI algorithm tasked with predicting future performance of a potential employee, student, or lender should prove that it is not basing its decisions on any of the protected characteristics.

In this case, one of the slogans of data-driven shortcuts (*"it is about knowing what, not knowing why"*) can no longer apply. How do you make sure that a data-driven agent like the ones described in the previous chapters does not end up learning that a certain protected group has a correlation with certain outcomes? Using this information would be unlawful, but it is difficult to audit when the decisions of the agent depend on statistical patterns discovered in terabytes of data, and implicitly represented by thousands of parameters.

LEAKING SENSITIVE INFORMATION

It may be difficult to check if statistical-AI agents make their decisions based on the right reasons, because they are designed to exploit correlations found in large datasets with no further consideration for causes or explanations. This is indeed the first half of the shortcut that gave us modern AI; the other half was that we can train these agents on data obtained "from the wild", that is, generated by social activities. This is where the machine could potentially learn some bad habits, as demonstrated by a remarkable study published in 2013 (which is so important to our topic that it will be discussed both in this and in the next chapter).

In 2012, a group of UK-based data scientists studied the Facebook profiles of 58,000 volunteers, accessing information such as age, gender, and religious and political views. They also inferred information about ethnicity and sexual orientation from indirect information that was present on the profile. At the same time, they also collected the set of "likes" on the Facebook pages of those individuals, and then asked a simple but unsettling question: can those sensitive and personal traits be automatically and accurately inferred on the basis of public statements such as "likes"?

The alarming answer is best summarised in the title of their paper: "*Private traits and attributes are predictable from digital records of human behaviour*". What they had found was that the set of likes of a user was sufficient to reveal their protected characteristics when compared and combined with information extracted from tens of thousands of other users. For example, older people were more likely to signal approval for "Cup of Joe for a Joe" or "Fly the American Flag", while younger people liked topics such as "293 things to do in class when you are bored". Males would prefer "Band of Brothers" while females liked "Shoedazzle". The power of the method came from combining several such "weak signals", which together produced a statistical signature sufficient to make good guesses about the protected traits of those individuals. The resulting statistical models were able to correctly distinguish between males and females (93%), homosexual and heterosexual in 88% of cases for men, 75% for women, African Americans and Caucasian Americans in 95% of cases, Christians and Muslims (82%), and between Democrat and Republican in 85% of cases, all on the basis of the "likes" published by each user.

The authors of the study observed that similar information may probably be extracted from other types of digital trails, creating a situation where sensitive information is leaked unintentionally. "*Similarity between Facebook Likes and other widespread kinds of digital records, such as browsing histories, search queries, or purchase histories suggests that the potential to reveal users' attributes is unlikely to be limited to Likes. Moreover, the wide variety of attributes predicted in this study indicates that, given appropriate training data, it may be possible to reveal other attributes as well*".

This important article will also be discussed in the next chapter since it also included predicting the psychometric traits of those same users. In both of these applications, the idea is the same: gather vast quantities of both private and public information about a large sample of users, and then use the data as a "Rosetta Stone" to infer private traits from public behaviour in different users.

How do we trust that an intelligent agent, trained to predict the criminal risk or future salary of an individual based on behavioural information, will not end up basing its decisions on unlawful information? This study shows that concealing the sensitive information from the agent might not be sufficient, because it is implicitly contained in innocent-looking and even public information.

The authors of the study concluded with a warning: "*Given the ever-increasing amount of digital traces people leave behind, it becomes difficult for individuals to control which of their attributes are being revealed. For example, merely avoiding explicitly homosexual content may be insufficient to prevent others from discovering one's sexual orientation*".

The "alien" constructs that a risk-assessment agent creates when it segments the training set of individuals, as a way to summarise internally the relations it detects, might well overlap with some protected information, without anyone noticing. Auditing this may be very difficult; more likely than not, those subsets of individuals will resemble the fantastic animal taxonomy invented by Jorge Luis Borges, which we described in Chapter 3.

Could protected information be inferred from our resume, even after redacting all the sensitive elements, and indirectly be used to make sensitive decisions in a regulated sector such as employment? This would be an example of an unlawful shortcut.

NEAR MISSES, FALSE ALARMS, AND SOME ACTUAL HARM

The use of intelligent algorithms creates the possibility to make statistical predictions about the future behaviour of an individual based on the past behaviour of millions of other individuals, and this can be used to estimate the risk of various decisions. This also creates the possibility that a machine can take a shortcut much like that of the monkey paw. This could be an unintended effect of combining the use of statistics, big datasets obtained from the wild, and unreadable representations. Finding reliable ways to audit these agents will be an important research direction to ensure trust.

Starting in 2016, the media has reported a number of close calls and false alarms. While it is not clear if anyone actually did get hurt in these media stories, it does seem clear that the risk assessment industry is looking at automating its decisions using statistical algorithms, and this is enough to cause concern about possible "monkey paws" and raise the issue of trust in algorithms. Legal regulation of this entire sector is necessary.

These are some of the stories reported over the past few years.

On October 11, 2018, the news agency Reuters reported on a story that we could consider a *"near miss"* since it never led to any harm. For several years (a research group within) Amazon had been developing an experimental software to review resumes and score job applicants. The article suggested that the algorithm was trained on a database of previous applicants and employees and was found to suffer from a serious problem: while learning word patterns in resumes that could be predictive of a good match, it "penalised resumes that included the word "women's," as in *"women's chess club captain* and *downgraded graduates of two all-women's colleges"*. The article missed too many technical details to help with our analysis, and Amazon never commented, except for clarifying that they never actually used that tool to evaluate candidates. If the report is confirmed, however, it does seem that the bias[1] must have entered the machine through the use of "wild data", which reflected the bias existing in the world. The concern is that another company, with less efficient internal processes, might be unable to audit their software and stop it on time.

On May 23, 2016, the online investigative journal ProPublica described a software tool that is used in some US courts to assess the likelihood that a defendant will become a recidivist. The software is called COMPAS (*Correctional Offender Management Profiling for Alternative Sanctions*) and has been used in various states, including New York, Wisconsin, California, and Florida to aid court decisions. Receiving a high score can have practical consequences for a defendant's liberty, for example, influencing their chances of being released "on parole", while awaiting trial. The article claimed that those scores were biased against African Americans, after comparing false positive and false negative rates in the two groups. Their conclusion was, *"blacks are almost twice as likely as whites to be labeled a higher risk but not actually re-offend"*, and *"whites (. . .) are much more likely than blacks to be labeled lower-risk but go on to commit other crimes"*.

This article provoked great alarm in the media, but not everyone agreed, and then it turned into a legal, political, and academic controversy; depending on how bias is defined and measured, different conclusions can be drawn. Importantly, while not releasing the code, the producers clarified that at no point does the method use information about the ethnic background of the defendant, nor any obvious proxies such as postcode. Their instrument asks 137 questions about the defendant, of which about 40 are combined in a formula to determine the risk of recidivism (the remaining questions are used for computing different scores). These cover the following areas: current age, age at first arrest, history of violence, vocational education level, and history of non-compliance. Could some of those behavioural signals have leaked the protected information into the algorithm? Later studies conducted by independent academics suggest that this was not the case. The main legacy of this controversy was to draw general attention to this very delicate area in which algorithms are already deployed: auditing these tools will be very difficult but also absolutely necessary for public trust.

Algorithmic translations can also be subject to unintended biases, as they rely on statistical signals extracted from natural data. At the time of writing (June 2022) Google Translate produced this translation, "*The president met the senator, while the nurse cured the doctor and the babysitter*" into Italian, "*Il presidente ha incontrato il senatore, mentre l'infermiera ha curato il medico e la babysitter*". This assumes that the nurse and babysitter are females, the president, senator and doctor are males.

In 2022, in the Netherlands, the Data Protection Authority fined the Tax Authority 3.7 million euros for using protected characteristics, such as ethnicity and origin, while scoring taxpayers for possible benefit frauds related to childcare allowance. The press reported that a software tool, defined by the investigators as a "self-learning algorithm", was used to highlight taxpayers for further examination, and it included ethnicity as one of the features used to assess the risk level of applications for benefits. Although technical details have not been released, this seems to be a case of what we call "algorithmic bias". The chairman of the data protection authority also claimed that some taxpayers were wrongly labelled as fraudsters as a consequence and denied access to opportunities; if confirmed, this would be a case where some harm would actually have been done.

BIAS FROM THE WILD

Inspecting a CV screening algorithm for bias is usually not as simple as checking which keywords it relies on, because the most recent versions of AI agents encode their knowledge in thousands or millions of parameters and use vast quantities of data to "tune" the formulas used for their decisions. This is one of the reasons why the makers of AlphaGo and GPT-3 cannot really predict or explain the behaviour of their creatures.

An ingenious study conducted in 2017 demonstrated how protected information could leak into the core of such algorithms, therefore providing the agent with an unintended signal that it could end up including in its decisions. This problem involves the very way in which words are represented within modern AI agents, creating the potential for unintended bias to make its way to the very core of an agent, and is described below.

Since the early 2010s, a way to represent the meaning of words in computers is by assigning them to points in a high dimensional space (usually over 300 dimensions) in such a way that proximity in that space reflects semantic similarity. This greatly helps any further analysis of a text but does create the problem of obtaining the coordinates of each word (the procedure is known as 'embedding'). Here is where machine learning meets "data from the wild" once more; we can extract high-quality embeddings for each word on the basis of the words that are frequently seen around it, which in turn requires processing large quantities of text. This powerful representation corresponds to an implementation of the ideas of British linguist J. R. Firth, who, in 1957, summarised his theories about meaning with the slogan, "*You shall know a word by the company it keeps*".

The specific way to obtain these coordinates has clear echoes of the battles of Frederick Jelinek at IBM in the 1980s, when he argued in favour of statistics over traditional linguistic rules. Today, most text analysis systems make use of word embeddings, trained on very large corpora that were sourced "from the wild". For example, GloVe (a widely used embedding of 2 million English words into 300 dimensions) was obtained from a set of millions of webpages containing a total of 840 billion words. These webpages included Wikipedia, many news sources, blogs, books, and various other types of data.

In 2017, Aylin Caliskan and her collaborators tested those very word embeddings for potential biases, finding that words such as job titles and job descriptions contained more information than expected; some were slightly shifted towards one side of the space associated with male concepts, some towards the opposite side, associated with female concepts. The most male words were: 'electrician', 'programmer', 'carpenter', 'plumber', 'engineer', and 'mechanic'; the most female ones were: 'hairdresser', 'paralegal', 'nutritionist', 'therapist', 'receptionist', 'librarian', 'hygienist', and 'nurse'. Those biases roughly followed the employment statistics of the general population. A similar finding could also be observed with academic disciplines, where arts and humanities are represented in a way that leans closer to a female sphere, and sciences and engineering closer to a male sphere.

How could this information have leaked into those coordinates? The only explanation is that in natural text, certain words describing jobs just appear more often in association with male words and others with female words—for example, pronouns. It is the use of data sourced from the wild that creates this bias. This explanation fits well with the documented gender bias found in newspapers by computational social scientists, who have analysed millions of newspaper articles, measuring gender bias in different topics and countries.

Once more the very shortcuts that gave the present form of AI seem to also have created the conditions that might lead to a monkey paw, for example if these embeddings were to be used as part of a resume screening software (I do not know of any such case being found).

THE MONKEY PAW AGAIN

The main point of this chapter is not the possibility of discrimination in the risk assessment industry, which is not even part of the core goals of AI, but rather the general possibility of a monkey paw by which a machine fulfils its tasks literally, indifferent to any harm it might cause along the way. The use of the data shortcut can definitely lead to that, which means that we need to develop ways to trust our intelligent agents before we devolve to them any amount of control over our lives.

Trusting someone (or something) means believing in their competence and benevolence, both of which need to be earned. The risk assessment industry, for example, might not have shown us enough in terms of its technical competence, but they have definitely shown us plenty in terms of what they intend to do with this technology. For example, it is difficult to miss that in the past few years, there have been several research papers containing the expressions "credit scoring", "social media", and "machine learning". This is because there is a potential business model in leveraging online behaviour to infer the credit risk of an individual. In separate but related news, in 2016, one of the main UK car insurance companies attempted to introduce a scheme that would set the price of car insurance based on social media information published by their owners. The scheme was scrapped a few hours before its official launch after Facebook clarified that their policy forbids using their data in this way. Now it is up to our lawmakers to make use of that information to regulate this important space.

As for the agents themselves, how do we trust that they are both accurate and respectful of our values? This is difficult when they operate in a domain for which there is no theory, so that they can only rely on statistical patterns from behavioural data. Currently, scientists and philosophers are exploring several dimensions of trust: transparency, fairness, accountability, accuracy, and auditability. By auditability, we mean that any software tool is conceived from the start for being easily audited by a third party, for example, an institutional body.

It might be a good idea to decide that only auditable technology can be used in regulated domains. This might create an obstacle to the use of certain representations of knowledge in machines, such as the word-embeddings described in the previous section, which behave much like the "alien" constructs of Borges: useful for the purpose of prediction, but not translatable into our language.

The founder of cybernetics, Norbert Wiener, in his prophetic 1950 book *The Human Use of Human Beings*, used the idea of the monkey paw to represent a device that pursues its given task in a "literal" manner, obeying its orders by taking an unintended path that no human would even consider. After 70 years, his warnings about devolving important decisions to intelligent machines sound incredibly relevant:

> *"The modern man (. . .) will accept the superior dexterity of the machine-made decisions without too much inquiry as to the motives and principles behind these. In doing so, he will put himself sooner or later in the position of the father in W. W. Jacobs'[2] The Monkey's Paw, who has wished for (two) hundred pounds, only to find at his door the agent of the company for which his son works, tendering him (two) hundred pounds as a consolation for his son's death at the factory".*

NOTES

1   Bias is a word that has different meanings in different fields, and we will use it to indicate a systematic deviation from uniformity or a norm. An algorithmic bias occurs when the algorithm makes decisions that are not uniform across different groups of users. In Chapter 7, we will discuss cognitive biases, intending them as deviations from idealised rational behaviour. In statistics, this word has different technical uses.

2   While Jacob's wrote two hundred pounds, Wiener wrote one hundred.

6

MICROTARGETING AND MASS PERSUASION

As we interact with online agents, we both leak and absorb a great deal of information, in a two-way relation that can shape our behaviour, as well as theirs. Both psychometric and persuasive technologies have been combined with this online interaction, and understanding how we can protect our right to autonomy will require significant research at the interface between many different disciplines, and eventually also legal regulation.

ONE DAY IN HAMBURG

When Alexander Nix addressed a transfixed crowd at the Hamburg "Online Marketing Rockstars" convention in March 3rd, 2017, he could not imagine how much his life was going to change in less than one year. Wearing a dark suit, thin tie, and

DOI: 10.1201/9781003335818-6

thick-framed glasses, he conveyed more of the image of Colin Firth in *A Single Man* rather than one of the "born-digital" enthusiasts forming his audience.

His presentation was entitled "From Mad Men to Maths Men", and it was about how data science and behavioural economics have transformed the business of online advertising. But the conference leaflet promised the story of how Mr. Nix's company had just helped Donald Trump win the election through "a shrewd microtargeting campaign" and "a workable model of persuasion". The company's name was Cambridge Analytica.

For 12 minutes, the audience waited patiently as Mr. Nix described the intersection between behavioural psychology, data science, and addressable-ads technology, three distinct methodologies that, taken together—he said—"are shaping the way that political campaigns are being delivered". Through that first part, all his examples came from product marketing.

"We rolled out an instrument to probe the traits that make up your personality", he announced three minutes into his presentation. "We use the cutting edge in experimental psychology: the OCEAN 5-factor personality model". He went on to explain that OCEAN stands for the five personality traits of openness, conscientiousness, extraversion, agreeableness, and neuroticism; then he described how different messages should be targeted to these different types. "Otherwise, you might end up sending the same messages to people with very different worldviews. For example, a conscientious person might be persuaded by a rational fact-based argument; (for) an extrovert, use language and imagery that evoke the emotion that buying a car would provide".

"It is personality that informs our decisions", he concluded, before introducing the idea of "addressable ads", the digital technology that allows the delivery of different messages to different people on social media. "Blanket advertising is dead", he declared, adding, "Our children will never understand that millions of people would receive one piece of communication.

We are moving to an age where (. . .) brands will communicate with you individually such that a husband and a wife in the same household are going to receive different communications from the same company about the same product".

None of that was news, and finally, after 12 minutes, the audience was rewarded for their patience; Nix started talking about the recent elections and the room fell silent.

"The work we undertook for the Trump campaign started in June 2016", he said, before explaining how they spent $100 million in digital ads and delivered 1.4 billion impressions in just four months, how they modelled the voters by focusing both on inferring their voting intentions and the issues that might motivate their choices, such as gun ownership or migration. "We assigned different issues to each adult in the entire United States," he concluded, noting also that all their work resulted in a net increase of 3% in favourability towards Trump.

At the very last minute, the screen behind Nix showed a final powerful slide: newspaper cuttings from the Daily Telegraph and the Wall Street Journal. The first was entitled, "The mind-reading software that could provide the secret sauce for Trump to win the White House". The second one had the subtitle, "Upset win was a coup for a tiny data firm that helped the campaign target ads based on a psychological approach". Nix modestly avoided saying that it was his psychometric targeting that was responsible for Trump's win, but his choice of slides did that for him.

During the following Q&A session, he was asked how he obtained the psychometric profiles used for targeting ads, and he explained that a large group of volunteers (in the hundreds of thousands) had taken an online personality test, creating sufficient data for an algorithm to find the relation between those personality traits and other data that was available about those users, which then allowed them to infer personality information for all other users. This specific detail would later become part of a controversy, but at that moment the audience was transfixed, and despite a few critics that questioned the ethics of working for a controversial candidate, the presentation was a success.

In March 2017, Alexander Nix was truly an "online marketing rockstar", but that would soon change. Just one year later, he would be explaining all that—and much more—to the UK Parliament, and briefly after that his company would go bankrupt. But that story is not of direct interest to us, as we focus only on the interactions between AI and society. So let us leave Mr Nix in Hamburg, standing in front of his slides, enjoying his success for a little longer, while we ask the questions that matter to us: Is mass-scale personalised micro-targeting, informed by personal data and performed by intelligent agents, possible? Can algorithms really find the best way to present the same choice to different voters? Can this method steer voter or consumer behaviour enough to make a difference in an election? How does this method work and who invented it?

THE ROAD TO HAMBURG

None of the ideas outlined by Alexander Nix originated with him; previous US campaigns had already made use of personalised targeting, linking individuals from public voters lists with commercial datasets of consumer behaviour, and using these to segment the voters into small homogeneous groups, a practice called micro-targeting. Also, social media had already been used before in US campaigning. As for psychographic segmentation, that was known to scientists too; two research papers, written before the 2016 campaign, explained how to infer personality information based on social media behaviour, and how to exploit the resulting information to improve "mass persuasion".[1] These two papers provide an important glimpse into the two-way interaction between human users and intelligent algorithms, an interaction that we urgently need to understand better.

Obtaining Psychometric Information from Online Behaviour

In 2013, the Proceedings of the National Academy of Sciences published a study demonstrating that a range of personal attributes of a Facebook user can be inferred based on their "likes" (this is the same study partly described in Chapter 5 about inferring "protected characteristics" of users). These attributes included psychometric constructs such as their personality traits.

The study started with a group of 58,000 volunteers, who had agreed to take an online personality test using a dedicated app on Facebook, and to share their user profiles and "likes" with the researchers so as to form a dataset which contained both psychometric and behavioural information for each user. The research question was: could the outcome of a standard personality test be predicted (i.e. statistically guessed) on the basis of the user's online behaviour? The remarkable answer was that this information, one of the key ingredients mentioned by Alexander Nix, can indeed be inferred from our online public behaviour, as we will see below. The implications of this finding for both the privacy and autonomy of social media users are still being digested today.

About Psychometrics

Psychometrics as a discipline aims to measure traits that cannot be directly observed, such as emotions, skills, aptitudes, attitudes, inclinations, and even beliefs and biases. These assessments are often done for recruitment, educational, or forensic purposes. It may sound surprising that a software agent could infer some psychological traits of a user, such as personality, by simply observing a sample of their behaviour, but this is how all psychometric testing is actually done; just typically these behavioural signals are constrained by a formal questionnaire or set of standardised tasks (often called an "*instrument*" in psychology). Psychometric traits are not observable, they are constructs aimed at summarising a subject's behaviour, that is, latent factors that have been postulated to account for patterns in the responses to a test.

For example, a standard questionnaire for personality types would ask you to agree or disagree with a long list of statements such as:

o "I am comfortable around people"
o "I get chores done right away"
o "My mood changes easily"

Statistical analysis of correlations within the answers and differences between different people can lead to models that attempt to explain them in terms of a few "traits" of these individuals. In the case of "personality", the theory most used by psychologists (from the 1980s) is called the Big-Five model, and it states that most differences in personality (and ultimately behaviour) can be explained by just five factors, which are remembered by the acronym OCEAN. These letters stand for: *Openness to Experience* (spanning a spectrum from routine-oriented to spontaneous and novelty-oriented), *Conscientiousness* (from disorganised and impulsive to disciplined and careful), *Agreeableness* (from cooperative and trusting to uncooperative and suspicious), *Extraversion* (from reserved to sociable), and *Neuroticism* (from anxious and prone to worrying to calm and optimistic).

Regardless of their "actual existence", those five scores can be inferred from past behaviour and are predictive of future behaviour; they are thought to summarise the thinking style and behaviour of a subject. They have been found to be fairly reliable across successive tests, relatively stable during a lifetime, to be shaped both by early experience and genetics, and to correlate with certain life outcomes, such as addiction. Of course, these scores cannot be estimated from single observations, and cannot predict individual decisions, but they correlate with statistical patterns of behaviour observed by aggregating large numbers of observations.

While a good "instrument" would repeat the same questions with different wordings, and take other measures to be accurate, there is no theoretical reason why one should not be able to use samples of behaviour obtained from "field observations": in principle, one can make psychometric assessments without a questionnaire, so long as one has access to samples of behaviour, for example, online behaviour.

As an example, psychologist James Pennebaker pioneered the study of a subject's choice of words to infer latent psychological traits or their emotional states, and Facebook filed for a patent in 2012 based on the concept of using a similar text-based method of inferring personality traits of their users from their online posts, so to use them for better recommendations of contents and adverts.

A separate paper in 2018 studied 683 patients of an emergency department—114 of whom had been diagnosed with depression—and analysed the words they used on Facebook in the months prior to the hospital visit to find that they could be used to identify the depressed patients better than random. So it is not too strange to imagine that also a public statement such as a "like" can convey some psychometric information.

The "like button" allows Facebook users to express their approval of various entities or ideas, for example, books, movies, and activities. Just this can say a lot about you; for example, liking a specific artist can reveal something about your age, ethnicity, values, and—most of all—aesthetic taste. And liking an activity such as cooking or reading can signal your preferences too.

The Psychometric Study

By analysing such public statements for 58,000 users, the authors of the psychometric study gathered enough statistical information to ask a cleverly posed question: would they be able to predict the outcome of a personality test if these users were to take one? This avoids all issues of "validation", that is, of proving that a test score does indeed carry information about a given trait, since the question is one of predicting the outcome of a standard test, and so any issue of validation can be transferred to that instrument.

They asked those users to take a standard personality test for the Big-Five traits, called "100-item International Personality Item Pool" (or IPIP) through a Facebook app, and then they compared the result with the set of likes they had published on their page. A machine learning algorithm was used to learn any correlations between these two separate sources of information, the "likes" and the personality traits, and these correlations were then tested on a separate set of users. Though not perfectly accurate, the personality scores predicted by the learning algorithm for "new" users were all significantly correlated with their actual test results. This important observation can be expressed in terms of the correlation coefficient, a statistical measure that has a value of 0 if the predicted test scores are completely uncorrelated to the actual ones, and 1 if they are perfectly correlated. For agreeableness, this quantity was 0.3 (where the test-retest accuracy for this quantity is 0.62, which we can consider as a baseline for how accurately this quantity can be estimated); extraversion had a correlation of 0.4 (compared with 0.75 test-retest accuracy); conscientiousness 0.29 (compared with 0.7), and openness 0.43 (compared with 0.55). Note that for openness, the algorithmic accuracy was close to the baseline test-retest accuracy of that psychometric indicator. In other words, online activities revealed information correlated with personality type.

For example, liking the singer Nicki Minaj was strongly correlated with being extroverted. Liking the character Hello Kitty was associated with openness. Using a separate psychometric test, the authors were also able to study and guess the IQ of users, with the unusual "curly fries" tag being one of the likes that tended to signal high IQ.

This study answers the first question posed by Alexander Nix's claims; it is indeed possible to know (something about) the personality type of social media users based on their public statements, such as their "likes". Presumably, one can even improve that by adding further lines of information or more training data. Of course, this does raise some problematic issues, such as that of "informed consent" by the user, but we will discuss them separately.

Note that this study has many connections with the ideas discussed earlier about GPT-3 and AlphaGo, particularly because it is another case where an AI agent has the opportunity to observe much more training data than a human can. A follow-up study conducted in 2015 compared the personality judgments of the algorithm with those of a user's friends, finding that the predictions of the machine were comparable and sometimes even better.

It is known that different people respond in different ways to the same stimulus, and this is the foundation of all forms of personalisation, from medicine to education to marketing. By segmenting a population into homogeneous subgroups, it is possible to target each segment differently; while traditionally the segmentation was done on a demographic basis (such as by age or gender), it is possible today to segment also on the basis of behaviour (e.g. based on interests or personal history).

The second scientific question raised by the claims of Mr Nix in Hamburg is: could marketing be more effective if different persuasive messages are delivered to different psychological types? In other words: since addressable ads (the direct delivery of specific messages to specific sets of online users) are a technical reality, are there any benefits in segmenting the user populations along psychometric lines? This is answered by a second study conducted by part of the same group before the 2016 US elections, but only published after that.

The Mass Persuasion Story

Online advertising is a game of large numbers, where often less than 0.1% of people engage with a promoted link, and where campaigns reach millions of people. The performance of a message is measured by "click-through rates" (the fraction of targeted users that click on the promoted link) and "conversion rates" (the fraction of those clicks that leads to some action, such as a sale), and even a small change in these base rates can make a big difference when scaled up to large audiences. As an extreme example, note that the 2000 US presidential election was ultimately determined by a margin of 537 votes in Florida. The game is mostly played by trying to match the best message with the best sub-audience in a way to have the highest success rates, a task that has rapidly changed from an art to a science over the past decades.

In 2014, a total of 3.5 million people were reached by ads as part of three separate online campaigns run by some of the same authors of the previous study, with the purpose of testing the effectiveness of different messages in different audiences. For our purposes, these can be considered as real experiments. The novelty of these three experiments was that the audiences had been segmented according to personality information, and different messages had been crafted to match the characteristics of each personality type. Would people respond differently to messages that are "congruent" to their personality type?

For that study, the researchers employed the same advertising platform that Facebook makes available for paying advertisers; while this platform does not allow audiences to be segmented by personality types, it does allow advertisers to address specific messages to a sub-audience based on the things they "like". In this way, by using the results discussed in the previous section, it was possible to identify and reach audiences that are statistically denser (or enriched) in specific personality types.

Experiment 1: Extraversion and Beauty Products

Over the course of seven days, different ads from the same beauty retailer were placed on the Facebook pages of female users. These messages came in various versions, tailored according to their degree of extraversion and introversion, as inferred from their "likes"; for example, extroverted users were identified through likes such as "dancing", while introverted ones by likes such as the TV show 'Stargate'. Messages designed for extroverted personality types included slogans such as "*Love the spotlight*", and those for the introverts included slogans such as "*Beauty does not have to shout*". In total, this campaign reached about three million users, attracting about ten thousand clicks and about 390 purchases on the retailer's website.

Having deliberately presented users with ads that both matched and mismatched their personality type, the researchers were able to observe that users exposed to "congruent ads" were 1.54 times more likely to make a purchase (conversion rate), when compared with those that were exposed to "incongruent ads", although there was no significant effect on the click-through rates. These effects may look like a small change, but they actually show an increase of 50% in the take-rate for congruent messages, were statistically significant, and were robust after controlling for age.

Experiment 2: Openness and Crossword Puzzles

The second campaign was aimed at persuading users to use a crossword app by tailoring persuasive messages according to their levels of openness to new experience, that is, the quality of preferring novelty over convention. It ran for 12 days on the advertising platforms of Facebook, Instagram, and Audience Networks. Like before, two audiences and two messages were identified, exposing both groups of users to both messages. The "likes" used to identify users deemed high in openness included

"Siddhartha", and those for users low in openness, included "Watching TV". A congruent message for a "high openness" user would have been, for example, "*Unleash your creativity and challenge your imagination with an unlimited number of crossword puzzles*" whereas one for a user that is low in that trait would have been, "*Settle in with an all-time favourite, the crossword puzzle that has challenged players for generations*".

The campaign reached 84,176 users, attracted 1,130 clicks, and resulted in 500 app installations. Averaged across the campaigns, users in the congruent conditions were 1.38 times more likely to click and 1.31 times more likely to install the app than users in the incongruent conditions. In other words, both clicks and conversions benefited from targeted ads. Again, the effects were robust after controlling for age, gender, and their interactions with ad personality; tailoring the message by personality resulted in over 30% increased engagement with the message.

The two experiments described above could fit with a philosophy of marketing as the task of finding the right customer for a given offer, so that it could be described as a service to the customer. But the third experiment came closer to what was claimed in Hamburg for the election campaign: given a set of voters and an intended outcome, the task was to find the best way to steer them towards the intended outcome. In this it resembles the example made by Nix when he said, "*a husband and a wife in the same household are going to receive different communications from the same company about the same product*". This is probably the setting that comes closer to the standard definition of "persuasion" (that will be discussed in further detail in Chapter 7, where we discuss behavioural economics).

Experiment 3: Introversion and Bubble-Shooter Games

The key feature of the third experiment is that it did not have the freedom to select the best users or products, only the best message.

The campaign was aimed at promoting a video game app of the bubble-shooter type on Facebook among users that were already connected to a given list of similar games. Since the list of users was pre-defined, and the intended outcome was also fixed, the only decision left for psychometric targeting was about which message would lead to the highest uptake. In other words, it was not about finding the best product for a user or interested users for a product, but just the best way to persuade a given user to make a pre-set decision.

The original generic message was *"Ready? FIRE! Grab the latest puzzle shooter now! Intense action and brain-bending puzzles!"*. An analysis of the "likes" of that fixed list of users revealed that the prevalent personality type was "introverted", so a new wording was created to match the psychological requirements of that audience. It said, *"Phew! Hard day? How about a puzzle to wind down with?"*

Both ads were run for seven days on Facebook, reaching over 500,000 users, attracting over 3,000 clicks, and resulting in over 1,800 app installations. The ads that were tailored to the psychological type obtained more clicks and installations than the standard ones. The click-through rate and the conversion rate were 1.3 and 1.2 times higher for "congruent" messages. Since both the audience and the outcome were predefined, the selection itself of a more congruent message increased engagements by 20%, providing probably the closest analogy to what would happen in an election campaign based on similar principles where a set of undecided voters in a hypothetical swing county has been identified, and then a congruent message is crafted based on their personality profile.

This can be considered as an example of what behavioural economists call a "nudge", that is, a change in the way a given choice is presented, but not in the choice itself or its economic incentives. This will be discussed in more detail in Chapter 7.

Taken together, these three experiments, which were published only in 2017, indicated that personality information inferred from online behaviour could be used to select the most persuasive messages, increasing engagement rates by about 30%. This is impressive if we consider the limited control that the experimenters had, only indirectly segmenting the audiences by specifying what items they "liked". A more deliberate and direct segmentation, perhaps based on a variety of behavioural signals, might increase these effects.

The resulting research paper that appeared only in 2017 summarised the lesson from the three experiments as follows, "... *the application of psychological targeting makes it possible to influence the behaviour of large groups of people by tailoring persuasive appeals to the psychological needs of the target audiences*".

THE ROSETTA STONE METHOD

The method used by the Cambridge researchers to infer personal information on the basis of public statements on Facebook is just an instance of a general approach to create the shortcuts that we discussed in Chapter 2. We will call it "the Rosetta Stone" approach, after the stone slab that carried three translations of the same decree by Ptolemy V, thereby enabling archaeologists to decipher hieroglyphics. Given different descriptions of the same information, for example, different views of the same Facebook user, we can use statistical algorithms to discover the relations between them so that we can translate further information from one representation to the other.

In this way we can connect the "likes" of Facebook users to their personality test scores and use this connection to predict the personality of new users. But we could also connect the demographics of a user to their purchases, or a text in a language to the same text in another language.

This becomes useful when one of the two representations is expensive, or not publicly accessible, and the other one is cheap, or public. In this case, one can infer the expensive attribute for a large number of people by simply observing the cheap or public ones, without the need to develop a deeper understanding than is strictly necessary—in other words, by taking a shortcut.

This idea has been used many times, for example to infer voting or purchasing intentions of a large population from a small number of telephone polls, when demographic details or purchasing history are available for all members of the larger population. The analysis might reveal, for example, that married homeowners who buy wine tend to lean Democrat or will buy sports shoes, allowing for more accurate targeting. Other known or attempted uses range from predicting credit scores based on social media content, to identifying customers who might be pregnant, to detecting potential frauds in the benefits system or users who might be about to switch to a competitor. All these inferences do not require the cooperation of the subject, which raises a number of ethical concerns, such as in a recent study that attempted to predict sexual orientation of social media users, and even some claims that lie detectors could be built along similar lines.

A traditional marketing statistician might say that this is just a case of "segmentation" of the population, whether this is done via demographics or purchasing behaviour or online statements. A trendier data scientist might say that those users that have shown interest in a given product (e.g. bought shoes) have been "cloned", or that "lookalikes" have been found. A key aspect of modern micro-segmentation based on online behaviour is that this is often based on very large amounts of very personal behavioural information, followed by machine-learning analysis.

One story has been particularly influential in popularising this approach, even though it turned out to be partly a myth. It still circulates at conferences and sales pitches as a folk tale to illustrate the power of data-driven marketing.

The Tale of the Pregnant Girl

In 2012, the New York Times reported that, many years earlier, data analysts at the retail chain Target had decided to send discount vouchers to pregnant customers, in order to secure their future loyalty. They had access to a large list of their customers and their purchasing history, created through loyalty cards and other means. In order to create a "Rosetta Stone", they needed a set of customers who were known to be pregnant, and this was assembled by using baby-shower registries and similar other methods that encouraged self-reporting.

According to that article, when jointly analysed, the two data-sets revealed that indeed certain specific purchasing behaviours can reveal when a customer is pregnant, and even her stage of gestation; unscented lotion is bought at the start of the second trimester of pregnancy, and *"sometime in the first 20 weeks, pregnant women loaded up on supplements like calcium, magnesium and zinc"*. The article reported that overall the analysts were *"able to identify about 25 products that, when analysed together, allowed him to assign each shopper a 'pregnancy prediction' score"* and also *"estimate her due date to within a small window, so Target could send coupons timed to very specific stages of her pregnancy"*.

The article included the entertaining tale of a father who had discovered that his daughter was pregnant only after receiving personalised coupons from the retail chain, and became rapidly popular. Of course, this version of the story is largely a myth, and—like in any good such story—vouchers were sent, the company ended up making a profit, and the analyst got a promotion.

An important ingredient in the above recipe is, of course, the public data relative to the entire population; how can a company access that if they are not Target nor Facebook? How could Cambridge Analytica acquire it in such a short time? It turns out that there is an entire ecosystem around the "Rosetta Stone" method, and this also includes the business of data brokers.

Data Brokers

"Want to know exactly how many Democratic-leaning Asian Americans making more than $30,000 live in the Austin, Texas, television market? Catalist, the Washington, DC, political data-mining shop, knows the answer". This was the memorable

opening line of an article published in Wired magazine by Garrett Graff in June 2008. The article was entitled "Predicting the Vote: Pollsters Identify Tiny Voting Blocs" and explained how the Democrats were investing in political micro-targeting companies. It added, "*They're documenting the political activity of every American 18 and older: where they registered to vote, how strongly they identify with a given party, what issues cause them to sign petitions or make donations*". It added in passing that the Republican party also had a corresponding project, and that such databases kept on growing, "*fed by more than 450 commercially and privately available data layers*".

This is indeed the main type of data that was also found in Cambridge Analytica's servers by the ICO investigation: publicly available datasets legally obtained from commercial data brokers, whose business it is to acquire, link, connect and sell data, for direct marketing purposes. The business of data brokers originated from traditional direct marketing and credit risk assessment companies, but it has now morphed into the collection, curation, and sale of personal data. They purchase information coming from multiple sources, such as loyalty card programs, as well as from the public record, and combine it. There are companies that boast having data for each individual adult in the US.

In the US, political campaigners can also access the Voters Registry; voters there are required to register to vote and, in many states, they have to declare their party affiliation if they want to take part in the primary elections for that party (which select the actual candidate). This voter information can be accessed by campaigns, and includes at a minimum the names, addresses, and party affiliation of individual voters. The combination of these data sources, with telephone polls about voting intentions and issues of interest, was used in 2008 by the Obama campaign and allowed the delivery of targeted messages.

The addressable ads offered by social media companies to advertisers provide a convenient way for them to target specific sub-audiences based on their behaviour, prompting Alexander Nix to tell the Hamburg audience, "Blanket advertising is dead".

ON HUMAN AUTONOMY

The two papers discussed above, on "mass persuasion" and on "private traits", are an exception rather than the rule, since most of these studies are conducted within commercial entities and do not need to get published. So these two studies are important because they provide a glimpse into the two-way interaction between human users and intelligent algorithms, an interaction that we urgently need to understand better. Taken together, those reports do answer the questions raised by the bold claims of Alexander Nix: it is indeed possible to both infer personality from online behaviour and use that information to inform the targeting of messages. We even know that the quality of algorithmic assessment of personality can be as good as assessments made by people.

They do leave us with a concern: if three academic researchers were able to do so much with so little, how much more can a company do with more data and more control? Indeed, the world of marketing has been revolutionised by AI, despite not being its core concern.

More research in this area is needed, as some of its consequences are not sufficiently understood. One reason for this is that it happens at the intersection between very different disciplines. It is this space that needs to be understood, as we investigate the two sides of the interaction: how a web interface can elicit personal information about a user, and how it can use that information to nudge the user in a given direction.

It is particularly important to consider the implications of Experiment 3 in the mass persuasion series, where information about personality is used to select the most effective message to persuade a specific individual to perform a pre-set task. This setting might be considered an example of manipulation more than a service. Importantly, the above psychometric methods do not require the collaboration or even the awareness of users, raising major concerns about consent.

Even more importantly, they raise concerns about human autonomy which is the capacity to make informed and uncoerced decisions, free from external manipulations, and which can be considered as a basic form of dignity, and as such should be part of the basic rights of individuals (at least in the European Union). Legal regulation of this area is overdue.

FULL CIRCLE

In May 2017, the UK Information Commissioner Office (ICO) started looking into how data was handled in that campaign, and in March 2018 the Guardian and the New York Times published an exposé alleging the use of personal data of Facebook users without consent. The affair became rapidly political, and more issues emerged during the investigation that eventually led to the demise of Cambridge Analytica. Those accusations, however, did not focus on the use of psychographic profiling for political targeting—which is still a legal practice—but on separate issues of informed consent, data sharing, and even entirely separate matters that emerged during the investigation.

The ICO inquiry included seizing 42 computers, 31 servers, 700 Tb of data, and 300,000 documents, all of which meant that conclusions were only reached in 2020. By that time, the ICO had uncovered rather unflattering details about the technology used by Cambridge Analytica. In October 2020, the ICO sent a letter to the Chair of a Select Committee in the House of Commons, reporting on their findings, which contained—among other things—the following observations:

> *"The conclusion of this work demonstrated that SCL were aggregating datasets from several commercial sources to make predictions on personal data for political alliance purposes".*[2]

(Point 9)

"SCL's own marketing material claimed they had 'Over 5,000 data points per individual on 230 million adult Americans.' However, based on what we found it appears that this may have been an exaggeration".

(Point 10)

"While the models showed some success in correctly predicting attributes on individuals whose data was used in the training of the model, the real-world accuracy of these predictions—when used on new individuals whose data had not been used in the generating of the models—was likely much lower. Through the ICO's analysis of internal company communications, the investigation identified there was a degree of scepticism within SCL as to the accuracy or reliability of the processing being undertaken. There appeared to be concern internally about the external messaging when set against the reality of their processing".

(Point 27)

In other words, the ICO letter said that the data they found was not particularly unique, and the internal staff of the company did not think that their prediction methods were all that effective. So, was Mr Nix's speech to the Online Marketing Rockstars itself a piece of marketing?

Of course, there was a lot more to this complex case that is not relevant to AI; there were issues of data privacy, informed consent, and even separate allegations of wrongdoing by the company which had nothing to do with technology. Some eyebrows were even raised by the academic paper on "private traits", with some questioning the possibility itself of giving informed consent in that study. But we will not enter into this, as it would take us far from our topic. When done with consent, psychometric profiling is still legal, as is the business of "data brokers", at least until further regulation is created.

As we said, the presentation by Alexander Nix in Hamburg was entitled "From Mad Men to Maths Men", a reference to the TV drama about a group of "creatives" at a New York advertising agency in the early 1960s. That title also echoed the battles between human experts and statistical algorithms that have taken place since the 1970s, when Frederick Jelinek first started exploring the data shortcut at IBM and joking about the benefits of firing human experts. That method, which we called "Rosetta Stone", applies to translations and product recommendations, so why not personalised delivery of ads? Whenever we have to make predictions in domains for which there is no theory, data-driven AI can be of help.

After the end of the affair, Alexander Nix understandably kept a low profile, briefly working for a new company linked to the same groups who had helped to fund Cambridge Analytica. These included a remarkable figure: the billionaire investor Robert Mercer, himself an influential figure in AI, who started his career at IBM in the 1970s as a close collaborator of Frederick Jelinek, precisely on the projects of speech and translation. In 2014, Mercer even received the same lifetime achievement award that had been given to Jelinek in 2009, for the same line of research on the application of data-driven methods to linguistics.

The statistical shortcut that enabled the psychometric approach to marketing and mass persuasion, and eventually led to that fateful day in Hamburg, can be traced back to those early ideas.

NOTES

1   Some of the authors of these articles worked in the same school of the University of Cambridge where some collaborators of Cambridge Analytica also worked, but these were completely different individuals, institutions, and projects.
2   SCL was the parent company of Cambridge Analytica.

7

THE FEEDBACK LOOP

The first intelligent agents with which we have a daily interaction are the recommender systems that populate our personalised news feeds on social media, a skill that they refine by constantly observing our choices and learning from them. We can regard them as our personal assistants, and empower them with our most personal information, or as control devices, whose task is to increase traffic to their web service, a goal that is not necessarily that of the user. The power difference between user and agent can be significant, considering that the agent can learn from the behaviour of billions of users, has access to personal information about each of them, and can select its suggestions from a near-infinite catalogue. We are just starting to understand the effects of this interaction on different levels: from steering clicking choices towards certain content to maximise engagement, to more permanent potential effects, which are not yet well understood, such as: self-control problems, excessive use, affective polarisation,

DOI: 10.1201/9781003335818-7

and the generation of echo-chambers where user perceptions are skewed. There are still no conclusive results about the effects of prolonged exposure of vulnerable individuals to this technology, and this should be an urgent concern

THE FIRST UBIQUITOUS AGENT

At about the time of the dot-com crash of 2000, the web took a turn towards "user-generated content", which in a few years led to the world of social media where we live today, in which everyone can publish or republish "content" easily, rapidly, for free, and often anonymously. For some this is now their main form of revenue; for many this is the main pathway to information or entertainment, including news, essays, podcasts, videos, music, and even entire books.

Social media is also one of the places where we encounter intelligent agents, to which we delegate the task of selecting the finite amount of content that we can actually consume. Sorting through the amount of media uploaded in just one hour is well beyond human capability, so we rely on the shortlists generated by the agent for the first screening: every time we access a service such as YouTube, Facebook, TikTok, or Twitter, we are presented with a personalised list of recent content that might be of interest to us, curated by their learning agent on the basis of what we, and millions of other people, did in the past.

While not part of its initial list of objectives, these systems are now one of the main revenue streams for the field of AI, as well as being essential to the current business model of most web companies, which depends on traffic to sell personalised advertising.

What are the consequences of long-term exposure to AI agents, tasked with the job of constantly finding something that we might want to read or watch or listen to?

Recent media reports suggest that the wellbeing of children and vulnerable users might be at risk, due to excessive or compulsive use, or to their emotional responses to recommended content. Some reports also suggest that our social wellbeing could be at risk, due to the potential polarising effect of echo chambers, a possible side effect of the algorithm whereby users would be automatically exposed to a stream of news that is increasingly skewed towards their beliefs and concerns. Rigorous scientific studies of these effects are still few and contradictory, despite the obvious importance and urgency of the problem.

We might have expected an automated doctor, or pilot, or maybe driver. Instead, the first intelligent agent that we have welcomed into our lives is in charge of selecting the news and entertainment for us. What do we know about it?

THE PERSONAL SHORTLISTING ASSISTANT

It turns out that we do not know much about the long-term effects of interacting with those agents, and the complete truth is that we are still struggling to find the right narrative to make sense of this "alien intelligence" that has become part of our lives. As we are faced with a new and complex phenomenon, we will take the advice of Richard Feynman, the Nobel Prize-winning physicist, who insisted that we should always try to look at the world in many different ways.

The most common narrative of the "recommender" is that of the intelligent assistant: an autonomous agent helping us with the hopeless task of choosing the best dishes from an infinite buffet. We could never make a rational choice, that is, one that has the highest utility, in a finite time, so the agent is designed to create a personalised shortlist from which we can then make our selection. In a situation where the cost of making a good selection is higher than the benefits we expect from it, it is actually rational to delegate the choice to an assistant, and this can well be an intelligent software agent.

Consider the model of a goal-driven agent that we introduced in the earlier chapters: it lives in (interacts with) an environment that it can partly—but not fully—observe and affect, and where it needs to pursue a given goal. This is done by selecting the most useful action in any given situation, a decision that is informed by whatever information the agent can obtain about the current state of the environment. In the absence of a reliable model of the environment, a very good alternative is to learn the best response for any given situation by experience and experimentation.

A recommender system does just that; its actions involve adding items to the shortlist that will be presented to the user and its reward will be the attention of the user, which will be indicated by some form of engagement. The available information will be observable properties of the available catalogue items and of the user, which are jointly called "signals". Over time, the agent learns how to recommend certain types of content to certain types of users to increase overall engagement.

There is nothing particularly surprising in the above description, except that there are about one billion videos on YouTube, with 300 further hours being uploaded every minute, and over two billion individual users. The signals that YouTube employs to describe its users and videos are in the dozens, and include which videos a user has watched before, what they have liked or commented on before, which ones were watched to completion, and so on; and for videos this will include captions, title, comments, reactions, date, and a lot more. Engagement is signalled by a combination of quantities, the most prominent one currently being "view-time" (but they also include sharing and comments). In this perspective, these quantities are proxies for other quantities that are not observable, such as interest, intent, attention, and satisfaction, and represent the reward that the agent craves.

The chances of an agent finding an engaging video for a generic user in these conditions would be quite slim, if it was not for an important factor: every day, five billion videos are watched on YouTube. This generates enough data for the agent to detect patterns, about which kind of users tend to engage with which kind of videos, and in this way starting the process of generating shortlists. Segmenting the users and the videos into homogeneous groups may be done implicitly, by a machine learning algorithm, but has the same benefits as the animal tax-onomies discussed in Chapter 3, as well as the same limitations, including that humans might not be able to understand the concepts created by the machine.

At a high level, the structure of recommenders is approximately the same for Facebook, Instagram, YouTube, TikTok, and all the others, though the details are different: which signals are used, and which combination of them is taken to represent "user engagement". Some of these may be kept secret, differ between platforms, and change over time within the same platform.

Before the final recommendation is made, the shortlists are further refined and personalised, based on the individual history of each user, removing various kinds of content and sometimes also adapting order and presentation. In the remaining milliseconds that they have left, they present you with the final shortlist.

And then they do something remarkable: they look at you.

BEHAVIOURAL ECONOMICS AND INFLUENCING BEHAVIOUR

We can also take the viewpoint of the company that operates the agent; for them, the goal is to attract visitors by providing them with appealing content so that each visit to the site is rewarding. They are not necessarily interested in providing assistance, if not as a means to a different end, which is to direct online traffic in specific directions.

This situation is analogous to that of another goal-driven agent: the flowering plant, which about 100 million years ago faced the problem of increasing its chances of pollination by attracting pollinating animals. That meant manipulating the behaviour of agents who had their own goals, which was done by evolving shapes, colours, odours, taste, and timing that were most effective in attracting—for example—insects. Benefits to the animal are possible, but are not the only way to steer their behaviour, and do not change the situation for the plant; its goal is just to maximise its own measure of "pollinator engagement".

If we consider this relation in economic terms, we can see at least three options for the agent: incentives, by which the controlled agent receives a benefit from the visit; nudges, by which the controlled agent is steered in a certain direction without any changes in the incentive structures; and deception, when one agent is misled by the other.

Behavioural economics, the study of consumer choices, can help us understand this relation. This discipline is interested in the way we make microscopic choices, especially when we deviate from rational behaviour. It has been known for a long time that the way a choice is presented—for example, to consumers—greatly influences the way they will decide.

While in Chapter 5 we defined "bias" as a systematic deviation from uniformity or a norm, here we consider "cognitive biases", intended as systematic deviations from an idealised rational behaviour.

Behavioural economists refer to the way a choice between alternative decisions is presented as the "choice architecture", which could include the order in which the options are listed, the wording used to describe them, and even images and colours used in the presentation. Steering the decisions of consumers by acting on the choice architecture without changing the actual options is called "nudging". This exploits cognitive biases that make our behaviour slightly irrational and are responsible for the fact that consumers are much more likely to spend £9.99 than £10 for the same item, or for our tendency to impulse-buy items at eye-level in the supermarket. Importantly, this is also responsible for the existence of "click-bait", that is, specific wordings in the titles of articles or videos that are effective in attracting clicks and are often manipulative or even misleading.

From the point of view of the flowering plant, it doesn't really matter if the pollinating insect receives any benefit in terms of nectar. The Australian hammer orchid flower mimics a female

Thynnid wasp, including a shiny head, furry body, and wasp pheromones. As the male wasp attempts to mate with those flowers, it spreads their pollen, without any practical benefits. In other words, the plant has evolved to hack its cognitive biases.

In a similar way, the recommender agent can learn which clues trigger our engagement by observing our choices, whether they are nudges or indicators of our actual interests. When this happens, do we reveal our preferences—as Nobel laureate Paul Samuelson would have put it—or rather our weaknesses, as a new generation of behavioural economists might prefer?

An example of how the most effective presentation of the same options can be learnt by an online agent is given by the legendary Silicon Valley tale known as "the 40 shades of blue".

In 2008, Google's graphic designers were working on developing a single style to use across all their products, and when it came to picking the specific shade of blue used by all products for hyperlinks, they had to make a decision. In particular, as a version of the legend goes, the issue was for hyperlinks leading to paid ads (on Google, advertisers only pay when they receive a click). Different designers had different opinions, so the executive Marissa Mayer (then VP of User Experience) made a decision: to test all 41 shades of blue—from the greener to the purpler—on random subsets of users to measure which design led to the most clicks. Then they waited. This is called an *a/b test* and is designed much like a clinical trial so that each group is identical to the others, except for one variable: the shade of blue. At the end of the experiment, there was a winner: a specific shade of blue that attracted the most clicks. The difference was not caused by different content but different presentation: a simple example of a nudge.

RECOMMENDERS AS CONTROL SYSTEMS

There is another way to describe the agents that recommend which content we might like to "consume". In this second narrative, all the recommender agent "wants" is to manipulate the behaviour of the user in order to get its coveted engagement signal, while all the company "wants" is to drive traffic first to their pages then to those of their advertisers. A way to understand this view of recommender systems is based on the ideas and language of control theory, which is the mathematical study of steering devices.

These are machines—called governors or controllers—whose task is to steer the state of another system, called the controlled system, towards a pre-set state, or maintaining it in that state. For example, the thermostat is a governor designed to keep your home's temperature within a given range. The key to its functioning is an information loop that constantly informs it of the current temperature so that it can apply the appropriate corrections. (The loop is formed when these corrections act on the temperature and their effect is measured). Similar systems are used to constantly adjust the altitude of a quadcopter drone, which would otherwise be unstable, or the velocity of a self-driving car. There is a unified mathematical language that is used to describe these systems and can be found across various domains, from economics to molecular biology. Today's control engineers use controllers that are both robust and adaptive, that is, they can function in uncertain environments and with incomplete information, learning on the job.

While these methods are more commonly studied for the control of physical systems, such as a heating plant or a helicopter drone, the same principles can also apply to the steering of goal-driven agents, so long as they react in a predictable way to the actions of the controller. Shepherd dogs can steer entire flocks of sheep because they instinctively know how individual sheep will react to their presence and barking, and central banks can steer markets by acting on interest rates.

It might be useful to regard Recommender Systems as adaptive "governors" designed to steer large numbers of human users by performing the right action at the right time for each user, much like the sheep dogs described above; they can sense the state of the user, through a multitude of signals, and select the most useful action from a vast inventory. Over long periods of time, they can learn how to best steer a user to increase their engagement. If we could replay this interaction at faster speed, it might be frightening to see how much the agent adapts to the user, and what long-term effect this has on the user's behaviour.

As an example, when YouTube changed its formula to define "engagement" in March 2012 from clicks to view-time (with the purpose of discouraging deceptive videos) they immediately observed a drop of 20% in their daily views (clicks on videos), accompanied by a steady climb of the average time spent by users on a given video, from one minute to four minutes. That year, the overall view time of the platform increased by 50%. Similarly, it was reported by the Wall Street Journal that an apparently innocuous change in the engagement formula of Facebook in 2018, intended to promote content that is likely to be shared with other users, might instead have led to an increase in content expressing or provoking outrage.

The way users respond to design changes in the recommender does fit with a view of these algorithms as control devices, and from a macroscopic perspective they do their job well. For example, the amount of time spent by the average US social media user reached three hours per day in 2018. Now, 80% of Americans report being online at least once per day, and 31% say they are online "almost constantly", although we cannot tell how much of this is a direct consequence of the recommender algorithm.

If we adopt this second perspective, then our micro-choices such as clicks and allocation of time are steered by the suggestions of an adaptive control system whose objective is to keep users swarming around its webpages, much like flowers are able to influence "pollinator engagement". In this sense, we are in the same position as the insects exploited by the plant for its pollination, possibly even in the position of that unfortunate Thynnid wasp.

Personalised Control Loops

While the "40 shades of blue" story demonstrates that it is possible to increase traffic by exploiting cognitive biases of the average user, an interesting question is whether this can also be done in a personalised way, that is, by discovering and exploiting specific inclinations of a given user. The question is particularly important since we know from Chapter 5 that much personal information is indirectly available to the agent, including of the psychometric type.

Indeed, this seems to be the case, as we have seen in Experiment 3 of the mass persuasion study described in Chapter 6. In that case, there was a given set of customers and a given product to recommend, so the only freedom the advertisers had was to select the best wording for each personality type. Since no economic incentives were changed, but only the presentation of the same choices, we should count that experiment as an example of a nudge that was selected based on psychometric information: a personalised nudge.

This raises an important question about the actual service performed by our automated assistants: whereas in general one could argue that they add value by finding the right products or the right customers for a given product, when these are both pre-decided all there is left for them to do is persuasion. In this case, are they serving or manipulating us?

An interesting story on this point was published by Netflix researchers in 2017, and it describes the benefits of automatically adapting the way an item is presented to a user, a final step carried out after the recommendation algorithm has compiled its personalised shortlist of titles.

Each title in the Netflix collection of movies has several different thumbnails, or artwork, that could be used to represent it on the user's menu, and different users could respond differently to different images. Since the infrastructure of Netflix can personalise the artwork that is used to represent each movie and record the choices of each user, the researchers were able to investigate how this choice can affect user engagement by using a machine learning algorithm to learn which images are more attractive to each user. The results showed that the click-through rate for titles with personalised artwork was higher than for those using randomly selected artwork, and also for those using artwork with higher average performance. In other words, personalising the presentation of the options, not the options themselves, led to a significant improvement.

While the 40 shades story presents a "collective nudge", and the mass persuasion story presents a personalised nudge hand-crafted by human experts, in the Netflix case, many of the decisions are automated. These cases are unlikely to be exceptions and they do indeed raise questions about the proper role of recommender systems. Today's content creators already strive to adapt their products to the preferences of the recommender agent. The day is not far away in which the content itself could be personally generated so as to obtain a required response from the user. The technology to generate images is already available, as demonstrated by the tool created by OpenAI and called DALL-E, which can generate high-quality images starting from a brief verbal description.

Legal scholar Karen Yeung has discussed the possible implications of personalised nudges delivered by an intelligent algorithm and called them "hyper nudges". As the agent has no way of distinguishing what represents our preferences and what reveals our weaknesses, could hyper nudges result spontaneously from the continuous feedback loop that is driven by the stream of user's choices on social media? I think they already do.

EFFECTS ON TWO LEVELS

Whether we frame the recommender system as an assistant helping to narrow down our options or as a control device trying to steer our actions, we should not be surprised when we observe that users' clicks are reliably influenced by the interaction. Recommendation systems do increase traffic, sometimes enormously, and they definitely increase the effectiveness of advertising: the combination of incentives and nudges that they can automatically discover is indeed intended for that effect, which we will call "the first order effect" and is described in the sections above.

What is not intended, however, is what we will call "*the second order effects*": by this we mean possible longer-term changes in the beliefs, emotions, or drives of a user, which the agent cannot directly measure and for which it receives no explicit reward. Can these occur unintentionally?

People are known to be affected by the media they consume, as many studies have found by interviewing viewers of different television channels, a phenomenon sometimes known as the "cultivation effect", through which not just their opinions but even their beliefs about reality are shaped by media content (e.g. famously our impression of the crime levels in our streets is influenced by the amount of television we watch). At the same time, repetitive and excessive exposure to a reward is known to create compulsive repetition of a behaviour that is associated with it.

While the first order effect of recommender systems is to steer the immediate clicking behaviour of a user, a second order effect may be to modify attitudes, interests, or beliefs of the user. Despite some anecdotal evidence and mounting concern, there is still no conclusive evidence of these unintended effects, but it is very important and urgent to study this question, due to its possible implications for individual and social wellbeing.

On Excessive Use

In 2019, Facebook published a study of 20,000 US users reporting that 3.1% of them felt they suffer from "problematic use", where this was defined as "negative impact on sleep, relationships, or work or school performance and feeling a difficulty of control over site use".

This excessive use could result both from first order effects (the nudges and incentives proving too powerful to resist for certain individuals) and from putative second order effects (e.g. if the user frequently returns to that service in a compulsive manner). There are various reasons to suspect that second order effects might be at play as we will see below (and as was suggested at a US Senate hearing in 2021 by Frances Haugen, a former Facebook employee, turned whistle-blower).

This perception is compounded by the claims of some entrepreneurs who—differently from clinical psychologists—can sometimes use the word "addiction" very freely. One company, that no longer exists, in 2017 advertised its services with the following pitch: *"(. . .) a tool that allows any app to become addictive, (. . .) people don't just love that burst of dopamine they get from a notification, it changes the wiring of the brain"*. This use of the word "dopamine" makes neuroscientists cringe, and is intended metaphorically, but it does reveal the mindset that went into certain design choices, and it does not help alleviate the concerns of parents of young social media users.

In a field study conducted in 2018 (often known as "the deactivation study") 2,573 US Facebook users were recruited and randomly split into a treatment and a control group. The treatment group was asked to abstain from using Facebook for four weeks. Among many other findings, the researchers reported that the usage of abstaining users was also reduced for weeks after the end of the period of "deactivation", concluding that social media is "habit-forming", which is another way of saying that current use increases the probability of future use.

Note that experts typically prefer to talk about "habit-forming" and "excessive use" because there are still no clear clinical criteria for a diagnosis of addiction in this case, as is the case for other behavioural addictions including gambling, sex, or compulsive shopping. More research will be necessary on the possibility of addiction to social media and on the role that recommender algorithms might play in that.

There is a lot we still do not know about the effect of long-term exposure to recommendation algorithms, particularly in children, given the reach of these tools in society. Even the 3.1% of problematic users, when applied to a user base of three billion people worldwide, would amount to 90 million individuals.

On Emotional Wellbeing

Various studies have reported an unexplained decline in emotional wellbeing in teenagers of various western countries since 2012. Some authors have blamed it on the simultaneous rise of social media; however, correlation is not the same as causation, and any association between social media usage and mental health outcomes would need to be explained by a convincing mechanism. Studies are underway to understand the effects of long-term exposure to social media and recommendation algorithms on mental health, including some large-scale field studies; some clues are already available.

The same "deactivation study" described above found that the group who abstained from social media had a small but significant improvement in self-reported wellbeing. A separate study was conducted at about the same time, with only one week restriction from Facebook on 1,765 US students, and obtained comparable results, namely significant but small improvements in emotional wellbeing among those who stopped using Facebook for a period, as well as indication of possible "habit formation".

One proposed mechanism involved the effect on mental health of negativity in news, coupled with the finding (observed on Twitter users in Spain) that negative content travels further on social media, so that one could expect that use of social media increases exposure to negative content and its psychological effects. This would also be compatible with the finding that increased use of social media does generally increase exposure to news.

A similar conjecture was presented in the Wall Street Journal during their coverage of the Frances Haugen leaks when a former Facebook employee handed over to US authorities a large set of internal documents. The journalists who saw those documents claimed that certain changes made to the recommendation algorithm in 2018 (particularly the engagement function), intended to increase the amount of content that is reshared by users, ended up increasing the amount of angry content instead. It is indeed conceivable that a learning algorithm, while attempting to locate the most "engaging" content, might instead end up promoting material that is richer in outrage and emotion, if this is what gets people's attention. Conclusive evidence, however, is difficult to find, also because the recommender algorithms are constantly changing.

The mental health effects of exposure to emotional content on social media have been studied by Facebook researchers in a truly interesting (if somewhat disturbing) paper that became known as "*the emotional contagion study*", which revealed previously unknown effects. That article also triggered various debates in academic circles, including an important one about "informed consent" by users, but here we will focus on its implications for the effect of recommender systems.

Over the course of a week in January 2012, about 650,000 Facebook users were randomly selected and split into three groups. While one of the groups would be left untouched to act as a control, the others had their newsfeed (the shortlist) slightly skewed towards more positive or more negative posts, defined on the basis of the words they contained (word lists are routinely used by psychologists to assess the emotional state of their subjects). During the same period, the posts produced by those users were also analysed in the same way. When the results came in, it was clear that people exposed to higher levels of positive emotion were more likely to post positive content too, while those exposed to negative emotion posted more negative content. The differences were small but significant. Importantly, the first group also posted less negative words and the second group less positive words, thereby suggesting that this was not a case of mimicry and likely reflected a change in the emotional state of the users. A further effect was observed: both groups who had increased exposure to emotional posts were more likely to post longer messages than the neutral control group.

The study demonstrates that even small changes in the recommender system could affect not just the amount of overall traffic in the platform (first order effects), but also the emotional condition of users (second order effects). The authors of that article summarised the significance of their study as follows: "*We show, via a massive (N = 689,003) experiment on Facebook, that emotional states can be transferred to others via emotional contagion, leading people to experience the same emotions without their awareness. We provide experimental evidence that emotional contagion occurs without direct interaction between people (exposure to a friend expressing an emotion is sufficient), and in the complete absence of nonverbal cues*".

The article correctly concluded that "*even small effects can have large aggregated consequences*" and pointed out that "*online messages influence our experience of emotions, which may affect a variety of offline behaviours*". The observation itself that exposure to news items with different emotional valence can lead users to write posts of different lengths is itself evidence of how even small nudges can affect our behaviour.

Mental health is particularly important when it comes to children, who are among the main users of certain social platforms. Ofcom (the UK authority that regulates telecommunications) reported that in 2020, half of UK children aged between 5 and 15 used social media, and this reached 87% for the age group between 12 and 15. About a third of children aged 5 to 15 used Instagram, Snapchat, or Facebook. It is clear that urgent work is necessary to establish the long-term effects of constant exposure to these algorithms on emotional wellbeing.

In an important ruling in September 2022, coroner Walker, Andrew found that recommender algorithms played a "more than minimal" role in the tragic death of Molly Russell, adding in his report: "The way that the platforms operated meant that Molly had access to images, video clips and text concerning or concerned with self-harm, suicide or that were otherwise negative or depressing in nature. The platform operated in such a way using algorithms as to result, in some circumstances, of binge periods of images, video-clips and text, some of which were selected and provided without Molly requesting them."

On Social Wellbeing

Another concern about the possible effects of long-term exposure to social media is that of political polarisation, that is, the emergence of increasingly extreme opinions or attitudes. This phenomenon has been observed over the past decade in the United States and other countries, particularly in the case of "affective polarisation" (increased negative perceptions of political adversaries), as opposed to ideological polarisation (an increase in the differences between policy positions).

One of the conjectures to explain this phenomenon involves the creation of "echo chambers", that is, subsets of users who consume a skewed diet of news that reflect, validate, and reinforce their political beliefs. These could be due either to "self-selection", which has been observed for decades in the choice of television news channels and whose effects are well documented, or to a phenomenon known as "filter bubble", that is, the tendency of a recommender engine to increasingly present contents that are similar to what the user has already consumed. In this case the expression "filter" refers to the recommendation agent, and "bubble" to a positive feedback dynamic that reinforces its own causes.

In other words, the circular effect of the recommender on user behaviour, and of user behaviour on the recommender, would create a runaway feedback loop which could lead to increasingly skewed perceptions, both in the user and in the agent, of their respective worlds. While the proposed mechanism is plausible and could also explain the increase in polarisation, there is still no conclusive evidence that this is indeed happening to a significant extent, but tantalising hints are becoming available.

A large-scale randomised field study (the gold standard in this kind of study) was conducted in 2018 by economist Ro'ee Levy, who created a cohort of 37,494 volunteers in the United States, randomly offering them subscriptions to conservative or liberal news outlets on Facebook. He also administered various opinion questionnaires and collected information about the news they consumed outside social media, among other things. This allowed him to compare the effects on users of exposure to news of similar or opposite leaning to theirs. The study established four facts: random variation in exposure to news on social media substantially affects the slant of news sites that individuals visit; exposure to news from the opposing political camp does decrease negative attitudes towards that camp (but no effect was observed on political opinions); and the recommender algorithm was less likely to supply users with posts from news outlets of the other political camp, even when the users had subscribed to that outlet (the subscription adds the news items to the longlist, but it is the algorithm that generates the shortlist).

Levy concludes by suggesting that social media algorithms might limit exposure to "counter-attitudinal" news and thus indirectly increase affective polarisation. As in other similar studies, these effects were statistically significant but small. On the other hand, the intervention applied by Levy was also small, being limited to just one subscription to a news source.

Another finding of the "deactivation study" discussed earlier (in the wellbeing section) was a reduction in the political polarisation of the group that had agreed to abstain from social media

for four weeks. This was measured again with various questionnaires, but in this case the reduction involved polarisation on
policy-issues, not affective polarisation. Again, the effect was
significant but small, and other studies failed to find a correlation at all.

Any connection between personalised recommendation and
polarisation is likely to be a complex one since affective polarisation has not been observed in some northern European countries
where the adoption of social media is comparable to that of the
United States, so it might depend on an interplay between the
technology, media landscape, and the political culture of a country. Once more, the most pressing questions point us towards the
interface between computational, human, and social sciences.

ON SECOND ORDER EFFECTS

Taken together, all these findings suggest that long-term exposure to recommender algorithms might really have unintended
effects on some users that go beyond the immediate steering of
their clickstream. As they affect those users' emotional states
and beliefs, these are longer-term changes that match our definition of second order effects. However, the observed effects
appear to be small *on average*, in terms of excessive use, emotional wellbeing, and social polarisation. It would be important to know if the small effect size is a result of measuring just
average effects, in this way potentially missing very large effects
in small subsets of vulnerable users; it is possible that different
people are affected in different ways from long-term exposure
to recommender algorithms in terms of excessive use and wellbeing. Another possible explanation for the small effect sizes
might be that the current studies are based on unrealistically
small "treatments", such as subscribing to just one more news
outlet or taking just a few weeks off social media. More research
will be needed to settle the question about unintended effects of
long-term exposure to recommendation agents.

FEEDBACK

So, are recommender systems our faithful assistants or our creepy controllers? We may want to keep both perspectives in mind while we find out more, as Richard Feynman recommended we do in the face of difficult questions. Indeed, some behaviours of those systems such as nudges would be difficult to explain within the narrative of the selfless assistant, as they add no extra benefit to the user, and they can probably be spontaneously discovered by the agent during its learning process.

It is understandable to be anxious when we devolve control. Recommendation agents can rely on super-human amounts of data including the distilled experience of billions of people, access to personal information, and powerful computing resources, so that the power difference between the human user and the digital agent might be significant. And yet, they have the limited wisdom of a statistical algorithm tasked with increasing "engagement" at all costs. There is the risk that they blindly pursue whichever measure of engagement they are given, usually a combination of clicks, views, hovers, shares, and comments, without regard to potential side effects on individuals or public opinion. Our concerns involve both the intended (first order) effects of the agents, that is, to steer the clicks of individuals, and the unintended (second order) side effects that we discussed.

Once more we can hear the distant echoes of the earlier Amabot battles of the 1990s, where human editors were replaced by data-driven algorithms. While the technique might be the same, this time the context has changed, with the news agents charged with curating an individual news feed for each user. It is possible that the original idea of Amazon's founder Jeff Bezos, of a different shop for every customer, is not appropriate when applied to news or even to content in general. Perhaps we should not want a different newspaper for every reader because we all need a shared reality, not one where we can cherry-pick news or—even worse—have it cherry-picked for us in a way to attract our attention.

Even if we did want a personalised selection of news, we might not want the selection to be driven by a need to "maximise user engagement", particularly if it emerges that negative emotion is more engaging than average and therefore is promoted by the algorithm, with effects on the wellbeing of both individuals and society.

What we need to understand is the role of human autonomy in this relation; is it ever acceptable to base a business on the notion of shaping the behaviour of millions of users? The possibility of hyper-nudging, that is, personalised deployment of nudges based on psychological biases, creates the possibility of an agent learning whatever specific combination of stimuli can get the attention of a given user, regardless of their actual interests or preferences.

So, we should be careful before we allow children and vulnerable adults to become part of a feedback loop where the other side is so powerful and not yet fully understood. It is surprising how little we still know about the effects of recommender systems on social and individual wellbeing. The choice to devolve control is ultimately a matter of trust.

As we deploy powerful learning algorithms in direct and constant contact with billions of people, we may realise that not only are those tools shaped by our behaviour but that we are affected by the relation too. The founder of media studies, Marshall McLuhan, was known for his memorable slogans, one of which was particularly successful: "We shape our tools and then our tools shape us".

8

THE GLITCH

Researchers have created intelligent agents that can learn a wide range of tasks just by experimentation, often to superhuman levels of performance. The example of an agent that plays video games provides a reminder that these machines can take unexpected shortcuts, exploiting properties of their environment that might be unknown to us, and failing to appreciate the broader meaning of their actions. How do we ensure that this does not happen with those agents that we have entrusted with delicate aspects of our lives?

STORY OF A GLITCH

In 2017, a group of researchers at the University of Freiburg conducted a study that consisted mostly of replicating existing results with a different technique, a routine but important part of the scientific process. In this case, however, the results were not what the scientists expected. To fully appreciate what they

DOI: 10.1201/9781003335818-8

discovered, we need to consider events that took place over 35 years earlier in the United States.

Q*bert

In 1982, the first video games were in great demand, not only in dedicated "amusement arcades" but increasingly also in homes, thanks to the newly invented video game consoles. So when Warren Davis and Jeff Lee created a new arcade game, it was natural that they would also want it to be adopted by the most popular console of the time: the Atari 2600.

This guaranteed an immediate celebrity status to the strange character of their game, an orange creature with no arms, a tubular nose, and a bad habit of swearing called Q*bert. The only thing stranger than that orange fellow was the environment in which it lived: a pyramidal stack of 28 cubes which changed colour when stamped upon. All Q*bert wanted from life was to change every cube into the same colour, by hopping diagonally from cube to cube, executing the right sequence of jumps without falling off of the stack. This feat gave it extra points and access to the coveted next level. Extra points would also grant it extra lives, which came in handy when a misstep made it plummet to its death.

Players loved it. Within a few years, Q*bert would have its own TV cartoon series and cameos in movies, no doubt thanks to the popularity gained by being available on the popular Atari 2600 console, the first mass-marketed device of its kind. It had been created in California in 1977 and included legendary games such as Pac-Man, Pong, and Lunar Lander. Other games were later ported to it, such as Space Invaders, Breakout, and—of course—Q*bert.

The 2600 was, in practice, a general-purpose gaming computer, able to support dozens of different games, which were contained in external cartridges with ROM chips; the users would buy a new cartridge, slot it into the console, and have a whole new game on their television. Against all odds, Q*bert's life and fortune would outlast even that of Atari.

Stella

The computer market changes fast and is prone to bubbles and bursts. In 1992, the Atari 2600, and other consoles of the same family, went officially off market after years of decline. Fans truly missed it, and years later, in 1995, Bradford Mott created an open-source version of the platform with all its games that he called Stella.

This was a clever piece of programming; Mott wrote software that accurately simulates the hardware of that old console in modern computers, thereby making it possible to execute all the original games in modern machines without having to reimplement any of them: it is a console without any hardware, entirely simulated in software.

In other words, by running Stella, any modern computer can pretend to be the Atari 2600, so that the very same original executable code of any old cartridge can be directly imported and run, as if it was still 1977. Nostalgic gamers also loved it because it was open source and free.

Arcade Learning Environment

More years passed and the world of computers kept on accelerating its constant transformation. In 2012, a group of AI researchers from Canada decided to use Stella as a testbed for their new machine learning agents; it naturally provided a complex environment where an agent could pursue a clear goal under controlled conditions. They called it the Arcade Learning Environment (ALE) and released it to the research community in 2013.

A new generation of machine learning algorithms could be compared in the same task-environments: Pac-Man, Space Invaders, Pong, Breakout, and—of course—also Q*bert. ALE rapidly became a standard testbed, and at age 30, well after its heyday as a TV star, Q*bert started a new chapter in its life.

DQN

Soon some of these researchers joined DeepMind, a company interested in developing more general forms of intelligent agents. Rather than developing specific agents to play individual games, the company wanted to develop a single intelligent agent that was general enough to learn any game of that type, and ALE was the ideal tool for that new quest.

Within a couple of years, DeepMind announced that they had created an agent that could learn any ALE game just by trial and error and *"surpass the performance of all previous algorithms and achieve a level comparable to that of a professional human games tester across a set of 49 games"*, using the same algorithm and initial settings for each of them. It combined two standard techniques that in AI are called "Q learning" and "Deep Networks", so that the agent was referred to as "Deep Q-Network" or DQN for short.

Learning a single game without any supervision is a remarkable feat in its own right; all the agent can see is the screen: an image of 210×160 RGB pixels, with a 128-colour palette, refreshed 60 times per second (60 Hz), as well as the current score of the game. Its available actions are (up to) nine alternative moves: the eight directions of the joystick and the no-move, as well as a "fire" button. Based just on this, and on the inbuilt drive to improve its score, the agent can learn by experimentation what are the right responses to any situation on the screen. It does help, of course, that the agent can rapidly play thousands of games in "fast forward", as it were.

The agent has no real idea of the meaning we attribute to those games, or the stories that we use to interpret them: the invading space aliens, the hungry little Pac-Man in the maze, the bouncing ball of Pong. DQN can just learn each of those 49 different tasks equally well, discovering rules, goals, and tricks at the same time, and without distinguishing among those categories.

In the specific case of Q*bert, the agent sees exactly what a human player can see: the full state of the pyramid and the score, and then it can move in four different directions. Points are awarded for each colour change, completing a whole stack, and defeating its enemies in various ways. Extra lives are awarded for reaching certain scores. But DQN does not produce the same narrative interpretations that we do; it describes the world in entirely "alien" terms to us and is indifferent to the drama of the little orange entity constantly desperate for more rewards.

The Freiburg Experiment

In 2017, a group of researchers at the University of Freiburg decided to try a different algorithm to repeat the same study on the same testbed. After all, this is what standardised tests are for.

The differences between their method and DQN are not important, compared to the surprise they had comparing its performance in the various games. There was an anomaly about Q*bert which they described as follows: " . . . *the agent discovers an in-game bug. First, it completes the first level and then starts to jump from platform to platform in what seems to be a random manner. For a reason unknown to us, the game does not advance to the second round but the platforms start to blink and the agent quickly gains a huge amount of points (close to 1 million for our episode time limit). Interestingly, the policy network is not always able to exploit this in-game bug and 22/30 of the evaluation runs (same network weights but different initial environment conditions) yield a low score*".

In other words, the machine had experimented extensively with the video game until it stumbled upon an unknown bug in its original implementation which increased the score by making some moves that—for us—are completely irrational. Why should Q*bert benefit from hopping on random cubes right after starting level 2? Why should it be rewarded for voluntarily plunging to its death from a specific cube at specific moments?

Why should this work only sometimes, but often enough to make a real difference?

What the Freiburg agent had discovered were just old "glitches", the likely result of programming mistakes or shortcuts taken at the time of implementation in the 1980s, that nobody had discovered until then. Its real power was not in its ability to digest more data than people can, but in considering actions that no human player would take seriously.

Despite the huge popularity of the game, nobody had experimented with it long enough or ever considered those highly unusual sequences of moves. But the bug had been there all along; after the publication, online discussion boards of hackers started comparing notes and experimenting with their old consoles, posting footage of the surprising behaviour being replicated on the original machine. The final consensus was that the glitch resulted from tricks that the original programmers Warren Davis and Jeff Lee had used to save memory, which was a very limited resource in the 1980s.

THE PERSUASION GAME

We may be tempted to describe the behaviour of the Freiburg agent as "irrational", but—according to our definition in Chapter 1—a rational agent is one that autonomously finds a way to maximise its utility, and this is just what that agent did; it discovered a useful feature of the environment that could be exploited, and it had no reason to consider it differently from all the other game moves it discovered, like hopping to dodge an enemy. From its point of view, we are the ones behaving irrationally when we do not take advantage of this property, but from our perspective, of course, its behaviour might well remind us of the monkey paw discussed in Chapter 6.

The distinction between a property of the environment and a glitch is only in our eyes. In science, we constantly look for "effects" to exploit, such as the tunnel effect in physics, and when they involve people, it is not clear if we should consider them as "features" or as "bugs". Take the "placebo effect", by which a patient can benefit from the (false) belief that a treatment was applied, it might be the result of a cognitive bias in the patient, but it could also be exploited to their benefit in the right setting.

Various cognitive biases in humans are known to psychologists, such as the "framing effects" whereby a subject's decisions depend on how a choice is presented to them, rather than on their actual consequences. We have seen how behavioural economics makes use of this cognitive bias to steer consumer choices, and how recommender systems are sometimes designed to automatically learn which presentation is more likely to nudge users towards certain actions.

The 40 shades of blue story of Chapter 7 is an example of a nudge that works well on average, while the Netflix artwork example shows nudges that need to be personalised due to the heterogeneous treatment effect. Between those two extremes, there is Experiment 3 in the mass persuasion paper (Chapter 6) where conversion rates of online ads can be increased by wording them in a way that matches the personality type of the recipient.

In each of these cases, an artificial agent finds the best way to induce the human user to perform a given action, exploiting biases that might be the result of ancient shortcuts taken by evolution or of our early experiences. We do not yet have a complete list of all such biases, let alone a good theory to explain them.

We could say that we, human users of recommender systems, are just one more gaming environment for agents of the same type as DQN or the Freiburg one: they all need to discover which actions will result in increasing the score at the particular game they are asked to play. In the case of the recommender, the environment happens to be formed or populated by human users and the reward function is their attention.

SERVANTS OR CONTROLLERS?

We could say that the comparison between content-recommendation and video games is inappropriate because the agent is there to simplify our access to the content we need and the interaction is what allows it to learn our needs. However, how do we explain the use of nudges in this perspective? Why would the agent need to manipulate the presentation of the same options in order to increase the chance that we engage?

An even more difficult situation to explain is the use of reinforcement learning, which is a technique intended to increase the long-term engagement of the user, rather than the immediate one; rather than making the suggestions that are most likely to result in the immediate consumption of content, which was plausibly what the user was looking for, a new class of agents can make suggestions that will result in increased engagement in the future. These agents were tested by YouTube in 2019, reporting small but significant improvements. Other companies are no doubt working in the same direction. It is worth noticing that the general structure of these systems was not dissimilar from the AlphaGo and DQN agents, being a combination of Q-learning and Deep Networks.

In the meantime, DeepMind has kept on improving the descendants of DQN, announcing in 2020 the creation of Agent57, reported to achieve "superhuman performance" on 57 different video games.

WHAT KEEPS ME UP AT NIGHT

When an agent has access to signals that no human can access, and to more experience than anyone can ever make in a lifetime, it does not need to be a genius in order to achieve "superhuman" performance at some task. A *savant* is more than sufficient. As there are billions of people that spend part of their day connected to these feedback loops mediated by AI, and some of those are our children, some obvious questions should be keeping us up at night.

Could there be specific nudges, possibly unknown because they are specific to small groups, that an AI can exploit as "glitches"? Maybe an innate attraction for the sensational, the emotional, the pornographic, or the conspiratorial? Maybe a personality type more inclined to compulsive behaviour, or risk taking, or paranoid thinking? Are users revealing their weaknesses to an intelligent agent that is indifferent to anything except for their engagement? Could a prolonged interaction with it lead some users to addiction, amplification, or other forms of problematic use? As a society, could extensive use of this technology lead to instabilities, such as polarisation or amplification of certain opinions, emotions, or interests?

After all, we do not need much research to know that we are a flawed species, that we often behave irrationally, both as individuals and as groups, and we also know from history that these flaws can be exploited to someone's advantage.

In these conditions, is it safe to couple ourselves with the same class of algorithms that figured out how to play all Atari games and Go better than people, in a game where the goal is to shape our attention and influence our behaviour?

As of 2022, the number of daily users of these tools is over three billion, and we still do not have conclusive scientific evidence about their long-term effects on vulnerable people. While I am inclined to believe that the level of concern in the media might well be exaggerated, we should properly study the effects on individuals and society that result from this new interaction, just as we would expect this from other industries. Fortunately, we are starting to move in this direction.

This is what keeps me up some nights: the idea of Q*bert hopping around in a meaningless frenzy so as to increase the utility of an algorithm. But what I actually see is a child, clicking around in a meaningless frenzy, just to increase the utility of an algorithm. It may be just one out of 49, like Q*bert, that has a vulnerability that we did not know about.

Over thirty years after its birth, after living so many different lives, Q*bert—the odd orange creature with a bad habit of swearing and a compulsion for hopping—might just have given its most valuable contribution yet: helping us reconsider what it means to allow an intelligent agent to play with our attention and shape our choices with the stated goal of increasing our engagement. More than W. W. Jacob's monkey paw, it could be Warren Davis and Jeff Lee's orange character, and what the agent from Freiburg did to it, that will represent the anxiety that we face.

9

SOCIAL MACHINES

As we search for intelligence, we may be looking in the wrong place: what if the intelligent agents that we encounter every day were actually social machines that include billions of participants? Depending on how we draw the boundaries of a system, we may see collective intelligence emerge from the behaviour of organisations such as ant colonies, and the same type of intelligence can be found in the massive social machines that we have created over the past two decades. Recognising them as intelligent agents may help us find better ways to interact with them.

THE ESP GAME

Two complete strangers face their screens at the same time; one of them is returning home on the night bus and the other one is getting ready for school. They were randomly selected from a large pool of online players and paired through the same app to play a game that will last only 2.5 minutes. They are thousands

DOI: 10.1201/9781003335818-9

of miles apart, but they cannot know this; they have no information about each other and no way to communicate. They probably would never give each other a second look if they ever crossed paths. Yet what they are asked to do is nothing less than reading each other's mind, and type the answer to this simple question: which word do they think the other one is typing right now? This is the only way to score points in the strangely addictive game they play.

The screen will not show a player all the content that the partner is typing, except for those words that they both have typed. With 273,000 words in the Oxford dictionary, what are the chances of them ever scoring a point in that short time?

This game might sound hopeless, except for an important detail: alongside the advice to "think like each other", they are also given a clue; they are both shown the same image, and they both know that.

As the timer counts down from 2.5 minutes, they frantically start typing what they think the other is typing based on that one clue. The best way to guess, it turns out, is to type anything that comes to mind by looking at that image, in the assumption that their partner will do the same. Which words would you try first, if the image was showing a black cat on a red mat and the clock was counting down?

As they scramble to read each other's mind, the players list whatever they think will come to mind to the other player in response to that image. The most obvious words, it turns out, are descriptors of the image itself; at least, these are the words that do match. If a certain image does not seem to yield many matches, they can both hit a pass button, and when they both do so, they can move to the next image.

The ESP game (short for extra sensorial perception) can be surprisingly intense and very addictive. But it has also an important side effect; it produces (and stores) very high-quality verbal descriptions of images, what we call "labels" or "tags". And this is a valuable commodity for learning algorithms that depend on labelled data for their training and testing.

In fact, this commodity is so valuable that in 2005, only one year after the game was created by student Luis von Ahn, the idea was bought by Google and used to annotate their images. The 2004 paper where this idea was first published estimated that in one month, 5,000 people can produce accurate labels for more than 400,000,000 images. We do not know how many images were labelled by Google from 2005 to 2011, when it was discontinued.

For all those years, the image search engine of Google benefited from these labels, and since these were later used to create the data to train modern image-recognition software, our AI is probably still using knowledge elicited from humans through this game.

GAMES WITH A PURPOSE

An important feature of this game is that it is impossible to spam: there is no shortage of merry pranksters on the web who would like nothing better than associating the name of some politician to a funny image. But doing this would require coordination between the two players, who were randomly selected and whose communications are tightly controlled by the interface.

And this is the key idea of the entire method: the only way to score points in these conditions is to reveal to the machine the information it wants. False labels would be readily revealed by a mismatch, and after an image has been used by just a few pairs of players, its key contents are reliably listed by the tags, so the algorithm can stop presenting it and focus on those images where there is still no consensus.

But there is another consequential point to this game: there is nothing that any player can do to stop the system from progressing towards its goal, which is to generate an increasingly refined and extensive annotation of a large dataset of images. This destination state, where each black cat is tagged both as black and as cat, is the purpose of the system, which is why this entire class of methods has been called "Games with a Purpose" (or GWAP) by its inventor, von Ahn.

Remarkably, this relentless pursuit of an innate purpose happens whether the participants want it or not, and even whether they are aware of it or not. Players do not need to realise that they are participants in a wider mechanism.

The ESP game showed that a system can be designed to serve two purposes, one at the level of the participant and one at a collective level, that do not need to be in agreement. It was also an example of "human computation", where the collective interaction among many people performs a computation of which they do not need to be aware. The dynamics of the game pushes the state of the dataset irreversibly towards a state of complete annotation, and this emerges spontaneously from the interaction.

Stafford Beer, the British cyberneticist who applied the ideas of control theory to human organisations, had a memorable quip about goal-driven systems such as this game: *"The purpose of a system is what it does"*. He intended to separate the intentions of those who created it, those who operate it, and those who use it from the intentions of the system itself. The notion itself of "unintended behaviour" refers to the intentions of users or creators but does not consider that at the level of the system, the goals might just emerge as the result of complex interaction. The principle is now known in management theory to describe the emergent results of self-interest of the individual participants in a social system, or of biological or other properties in a technological or a natural system. It even has its own name, based on the initials of the quip: the POSIWID principle.

The idea of designing distributed computations through the interactions of human participants leads to a remarkable inversion of roles; the participants assume the role of "cogs" in a wider machine whose ultimate purpose they cannot see, whereas all the consequential decisions emerge at the level of the machine, which is the one pursuing a goal and possibly exercising autonomy.

The mathematical theory telling self-interested players how to best play a given game is called "game theory", and it will readily explain that in the ESP case, the optimal strategy is to describe the image in simple terms. However, there is also another theory telling game designers how to best steer the self-interested behaviour of participants towards their own system-level goals by designing the rules of the game; this is called "mechanism design" and is one of the most powerful parts of applied mathematics.

Mechanism design sets the rules of auctions, where all the participants have the goal of saving money but the auctioneer has the goal of extracting the highest price for any given item. The mechanism is designed in such a way that no participant can pursue their true goal without disclosing part of the information they are interested in concealing, until the system identifies the participant who is willing to pay the most and what is the most that they would be prepared to pay. That is the purpose of the system, because that is what it invariably does.

Markets are very much like the ESP game, where the price of a good reflects what every participant thinks that the other participants will be prepared to pay for it. Besides being a mind-reading game, markets can also process information, reflecting the consensus prediction of thousands of players, who might be using different information sources and modelling strategies. While it would be difficult for traders to guess if others will buy or sell a given stock tomorrow, and therefore place a bet, they can do this because they are all looking at the same economy. As a result, the system can maintain an updated estimate of the worth and prospects of different parts of the economy, at least on a good day when it is not affected by pathological phenomena such as market bubbles.

The Invisible Hand

In the context of markets, the spontaneous emergence of purposeful behaviour from the interaction of self-interested participants has a name: the "invisible hand". This comes from a fortunate definition by political philosopher Adam Smith, who wrote in *The Wealth of Nations* that "*Every individual (. . .) intends only his own gain, and he is in this, as in many other cases, led by an invisible hand to promote an end which was no part of his intention*".

The next time you use eBay, the global auction website, think that you are a participant in a large game with a purpose, where all other participants can see the same goods and bid for them. While their individual purpose is to spend as little money as possible, concealing their maximum spending limit, the purpose of the overall system is the exact opposite: to identify those users willing to spend the most. They are part of a machine—not one made of metal, or electrons, or cells, but one made of people. Here the goals of the participants and those of the system they are part of are spectacularly misaligned, but there is nothing any of them can do about it; the only way for each agent to pursue their local goal is to further the ultimate goal of the overall system.

On Alignment

The goals of participants can be fully aligned, as in the ESP game where players must cooperate, or fully misaligned, as in eBay where bidders must compete. But what we should not miss is the alignment with the goals of the system; in the case of ESP, most participants would be indifferent to the goal of annotating images, whereas in the case of an auction, the goals of the auctioneers and those of the bidders are opposite. There are also cases in between. In all these cases, however, the common feature is that the participants cannot control the overall system nor can they withhold information if they want to pursue their own local goals. Participants became parts of the machine, without needing to be aware of it, and without being able to control it.

SOCIAL MACHINES

The ESP game is an example of a "social machine", and you do not need to play an online game in order to become part of one.

A machine is a system formed by several interacting parts, each with a separate function, which together perform a particular task. Machines can be made of mechanical, electrical, hydraulic, and even chemical or biological parts. An old-fashioned camera would have optical and chemical components, a windmill would have hydraulic and mechanical components, and a fermentation plant for the production of biofuel would also include biological elements. So why not have human components too?

Cybernetics, the discipline that in many ways underlies today's AI revolution, had as its foundation one idea: that the same principles must apply to mechanisms, organisms, and social organisations. For example, the seminal book by Norbert Wiener that created the field was named *Cybernetics: Or Control and Communication in the Animal and the Machine* and the conference series that popularised the field was called *Cybernetics: Circular causal and feedback mechanisms in biological and social systems*. In that theory of machines, it was only the behaviour of the components and how they were interconnected that mattered, but not the details of their nature. There is no reason why people could not be part of machines too.

You can imagine—in a mind experiment—replacing each component of a complex apparatus with human operators who follow detailed instructions, and its overall behaviour—in principle—would be the same. For example, counting ballots, adding numbers, sorting the mail, or routing telephone calls: all tasks that were initially performed by people tightly controlled by a rigid infrastructure.

We call a "social machine" any system that includes humans, each performing well-defined and narrow tasks and whose interaction is mediated and constrained by a rigid infrastructure. Today, this infrastructure is typically digital, but does not need to be: a physical bureaucracy communicating via standardised forms or a moving assembly line could be regarded as social machines in which the human participants do not need to be aware of the overall objectives of the system, as they are only required to complete narrow and local tasks in standardised fashion. In fact, a set of people interacting via standardised forms and following detailed instructions could perform any computation. Alan Turing called the combination of a human following a detailed set of instructions by hand a "paper computer".

We will focus on social machines that are mediated by web-based interfaces and we will call the people that form them "participants". The term "social machine" was introduced in 1999 by Tim Berners-Lee, the inventor of the World Wide Web, who used it in a rather more optimistic way, when he wrote: "Computers can help if we use them to create abstract social machines on the Web: processes in which the people do the creative work and the machine does the administration". We will use his terminology, though many of the examples that we encounter today seem to suggest a rather concerning reversal of the roles, with human beings doing the routine chores and the digital system setting the overall objectives.

The function of the digital infrastructure as a mediator between all participants, is not just to coordinate the work that each of them contributes but also to allocate incentives, since these components of the machine are themselves autonomous agents who voluntarily choose to participate because they expect a benefit. One of the jobs of the machine is to maintain itself, by recruiting and rewarding participants.

For example, we could imagine a variant of the ESP game that also distributes some valuable asset, for example information or entertainment, to those players who meet some targets. This aspect is essential in modern social platforms, which depend on the presence of billions of voluntary participants, so that incentives must be delivered, also on a personalised basis, which in turns requires a high level of information elicitation. We could even imagine a variant that locates the players that are most likely to break a tie between two competing tags, and the best incentives to secure their participation. Most of the activity of the social machine might well be spent preserving itself and its internal consistency.

This would be a familiar concept for cybernetics researchers of the 1960s, who liked to talk about living systems in terms of "*autopoiesis*", the quality of being able to create themselves, and "*homeostasis*", being able to maintain a stable internal environment.

Since both the machine and the participant can be agents, it will help the clarity of our discussion if we distinguish between those two levels, which we will call the macro and the micro level. Actions and goals can be very different at different levels, as can the amount of information available for the necessary decisions. A movie or music playlist is generated by a social machine at the macro level, while the decision by the user to like or share a specific videoclip is a decision at the micro level. The annotation of images generated by the ESP game is an emergent property at the macro level, the specific choices of which words to type are decisions at the micro level. While it is clear that participants are autonomous and goal-driven, it might not be so clear that the macro-level agent may be too, and that will be the topic of the next section.

The idea of a social machine provides a useful abstraction to think about the vast and complex socio-technical systems that have emerged on the web, such as the social networking platforms and the search engines, some of which have billions of active users on any given day. Not only are eBay and the ESP games social machines, so are the online communities that collectively vote for the most useful answers to a question or the most interesting videos, though their different rules result in different emergent behaviours. In those cases, a web interface shapes all the interactions among the human participants, constraining both the information and the options that they have, so that they can only pursue their objectives by also furthering those of the overall machine.

AUTONOMOUS SOCIAL MACHINES

The examples we have made of web-based social machines have the extra property of not being controlled by any external agent: their overall behaviour emerges spontaneously from the interactions between their participants, as well as those with the environment. While autonomy is not a necessary requirement for a social machine, it is of great interest in our study of autonomous agents.

We will focus on web-based autonomous social machines that incorporate millions or billions of participants and are mediated by a learning infrastructure, such as the one found in recommender systems, so that they can leverage and generalise the data gathered from all of their participants. The amount of information storage, the rapidity of communication, and the flexibility of the interaction are unlike any other previous attempt at making social machines.

The question we will ask is: can certain social machines be regarded as intelligent agents?

GOAL DRIVEN SOCIAL MACHINES

In Chapter 1, we defined an intelligent agent as one that can pursue its goals under a range of conditions, including also situations never experienced before, in a complex environment. Such an agent should be able to gather information, use it to make a decision, and possibly also adapt to the environment so as to behave better in the future.

We do expect organisms and algorithms to be capable of this kind of behaviour, but even organisations can act in this way: we do know that ant-colonies (or other collective entities) are capable of making very complex decisions where no individual ant has sufficient information—nor brain power—to understand what the whole colony needs to do next.

There is no central commander nor central plan in an ant colony, yet each ant knows what to do next, in such a way that the colony responds to changes in the environment. As new food becomes available, or the territory of the colony shrinks or expands, the individual ants can change their role—say from housekeeper to forager—or adjust their patrolling path to fit the new situation. These decisions require global information that is not accessible to the individual ant, such as how much food the others have discovered or collected, and therefore how many foragers are needed.

This goal-driven behaviour at the macro level emerges spontaneously from a different behaviour at the micro level: individual ants might collect information about how often they bump into each other or about chemical markers left by the scout ants when they find food. Patrollers attract each other to the food they have discovered until a global decision is made about the most promising site, and later foragers go where most patrollers are.

Among the many complex decisions a colony must make collectively, one is when and where to build a new nest, and some ant species can even build bridges and rafts in case of flooding. All these are examples of goal-driven behaviour, collectively made by an organisation, without any individual being in charge. Sometimes this is called "emergent behaviour" and even "collective intelligence", and it is reminiscent of the "invisible hand" of Adam Smith.

As discussed in Chapter 1, philosophers sometimes call "*telos*" the spontaneous direction of movement of a system, a Greek word used by Aristotle to denote the ultimate goal or purpose of an activity or agent, so that goal-driven machines are often also called "teleological".

While auctions and the ESP game are explicitly designed to ensure that the macroscopic goals are always pursued, there is no known theoretical guarantee in the case of emergent behaviour in ant colonies, other than that their behaviour has been moulded by millions of years of evolution. Freeloaders and saboteurs are theoretically possible in all social machines, except for the simplest ones.

RECOMMENDER SYSTEMS AS SOCIAL MACHINES

We all become part of a social machine as soon as we use a recommender system, which means several times per day. When we log into YouTube, we are greeted with a personalised shortlist of suggestions, and as soon as we select one, we contribute to the annotation of its massive catalogue of videos and users. Just as the price of items on eBay reflects the activities of the participants to that social machine, and the tags annotating the images of the ESP game reflect a general consensus among its participants, so also descriptions of content and users emerge directly from the usage of these systems and inform their future behaviour.

It is not easy to draw the boundaries of any system, but we could say that the behaviour of YouTube is determined by the collective activities of its billions of users.

The general idea of collectively annotating content through a social process can be traced back to 1992 when Dave Goldberg, computer scientist at Xerox PARC (the legendary Palo Alto Research Centre of Xerox), invented a way to filter emails and newsgroups based on how other users had tagged them, an approach that he called "collaborative filtering". In the same article where this idea was presented, he also introduced the term "implicit feedback" to describe an annotation of the data that is obtained by simply observing the choices of users.

The difference between collaborative filtering and eBay is simply in the incentive structure: while the same web post can be read by everyone, a given item on eBay can only be bought by one user; this creates incentives for collaboration in the first case and for competition in the second. Valuable items end up having more readers in the first case and a higher price in the second.

Today we leverage micro-decisions made by millions of participants in order to recommend books, videos, news, and emails. The agents behind those decisions are ultimately collective entities, tasked with discovering and evaluating resources much like the ants in a colony do for food. It would be interesting to see if these systems can detect changes in their environment that we—the participants—cannot notice or comprehend. Considering the amount of data those macro agents can access and process, could they become an example of super-human intelligence?

Goldberg's definition of "collaborative filtering" referred to the relation among users, but we should remember that the goals of participants and those of the macro system may be very different, to the point that we might want to consider the possibility of "competitive filtering" too. In order to decide if we are in a competitive or collaborative relation with a recommender system, we should be able to observe the goals of the macroscopic system. Philosophers call this discrepancy the problem of "value alignment".

What does YouTube want? This may change, but at this moment it seems to be to increase the total platform-view-time, or some similar measure of engagement. One may also think that it wants to maintain a sufficient number of participants so as to maintain itself. The purpose of a system is what it does, not the intent of those who operate it, nor that of those who use it.

The telos of a recommender system emerges from the interactions of millions of users, but it is also shaped by the learning algorithm that in turn is driven by the formula used to calculate the reward that it receives for the reactions of the users, so that the invisible hand can be partly controlled by acting on that parameter, but there is no mathematical guarantee that the entire system will follow meekly. Maybe the right way to see the recommender system is as a controller of an entire and reluctant social machine.

REPUTATION MANAGEMENT SYSTEMS AND ALGORITHMIC REGULATION

Recommender systems are not the only social machines in widespread use today; more examples are the reputation management systems employed, for example, to score both customers and providers of services such as car or house sharing. In this case, the feedback is not implicit, but rather entered in the form of a satisfaction rating. We encounter them every time we book a hotel room online.

In 2013, the Silicon Valley futurist Tim O'Reilly proposed that society might be better regulated via feedback loops than via top-down law enforcement. His example was the way in which Uber, the ride-sharing app, can regulate the behaviour of both drivers and passengers by using a reputation management system:

> *"[Uber and Hailo] ask every passenger to rate their driver (and drivers to rate their passenger). Drivers who provide poor service are eliminated. As users of these services can attest, reputation does a better job of ensuring a superb customer experience than any amount of government regulation (. . .)".*

What he was describing was the application to governance problems of a technology that has become standard in the private sector: the compilation of scores, about businesses or even individual professionals, on the basis of customer feedback. For example, after using a restaurant, hotel, or even receiving a parcel, we might be asked to fill out a satisfaction rating form. The aggregation of those ratings results in a score that is considered a proxy for the reputation of a business, or person, with the intention to shape the choices of prospective customers.

This is a familiar example of a social machine in a very consequential position for many traders, which does not, however, come with the same mathematical properties of eBay or ESP since it is conceivable that the vote of a customer might be influenced by the current score or other factors rather than truly revealing the quality of their experience. As for recommender systems, so also for reputation management systems: there is no mathematical guarantee that the system is robust to attempts to skew its "beliefs", and indeed both spamming and echo chambers are possible in those systems.

Could this really be a model for social governance too, as suggested by O'Reilly? Some countries are experimenting with the idea of assigning a "*social score*" to their citizens; however, that would be determined by their actual behaviour rather than the feedback from their peers. Connecting opportunities (such as education) to that score could be a powerful incentive to adapt one's behaviour, although it is not clear how this would steer a whole society. Certain companies also experiment with scoring their workers based on performance and even on reputation.

We should keep in mind that the ultimate "purpose" of a system may be different than that of its operators, and nonlinear phenomena such as feedback loops can easily affect both recommendation and reputation systems, just as they also affect markets and social networks. If we cannot stop fake news or investment bubbles, why should we expect to be able to manage algorithmic scoring services? Besides these technical considerations, there is also the ethical consideration that just by making use of any reputation management system, we create pressure for others to join it and increase the power of the few who run it to shape the behaviour of the many who depend on it. In other words, this might create a new powerful social mediator that could potentially have harmful consequences.

THE STORY OF AUNT HILLARY

In the marvellous classic book *Godel, Escher, Bach* by Douglas Hofstadter, an anteater tells the story of the eccentric Aunt Hillary, who is a conscious ant colony.

"Ant colonies don't converse out loud, but in writing", he says, explaining that the trails formed by the ants can be read like words, as they change shape in response to his questions. *"I take a stick and draw trails in the ground, and watch the ants follow my trails. (. . .) When the trail is completed I know what Aunt Hillary is thinking and in turn I make my reply"*.

When his friends are surprised that he—an anteater—should be friends with ants, he clarifies, *"I am on the best of terms with ant colonies. It's just ants that I eat, not colonies-and that is good for both parties: me, and the colony"*. And he explains that his job is that of a "colony surgeon": by carefully removing individual ants, he helps the health of the colony.

The anteater insists not to confuse the two levels; it is the colony that he converses with, the ants would not be able to understand him. *"You must not take an ant for the colony. You see, all the ants in Aunt Hillary are as dumb as can be"*.

And then he describes how Aunt Hillary greets him upon his arrival, by changing her footprint on the ground.

"When I, an anteater, arrive to pay a visit to Aunt Hillary, all the foolish ants, upon sniffing my odour, go into a panic-which means, of course, that they begin running around completely differently from the way they were before I arrived (. . .) I am Aunt Hillary's favourite companion. And Aunt Hillary is my favourite aunt. I grant you, I'm quite feared by all the individual ants in the colony-but that's another matter entirely. In any case, you see that the ants' action in response to my arrival completely changes the internal distribution of ants. The new distribution reflects my presence. One can describe the change from old state to new as having added a "piece of knowledge" to the colony".

To his surprised friends, the anteater then explains the key to understanding the two levels (which we call micro and macro, or the participants and the machine):

> *"If you continue to think in terms of the lower levels-individual ants, then you miss the forest for the trees. That's just too microscopic a level, and when you think microscopically, you're bound to miss some large-scale features".*

Finally, after explaining in further detail how the colony can process complex information that individual ants cannot understand, the Anteater concludes, *"Despite appearances, the ants are not the most important feature"*. Later on, he also adds, *"It is quite reasonable to speak of the full system as the agent"*.

We can see the similarity between Aunt Hillary and a modern recommender system, where a multitude of individual users just contribute to its behaviour without having any real control or understanding. As we face the welcome page of a social media website, we deal with the macroscopic agent, and we do not need to know anything about the individual participants. Like the anteater might get along with the macroscopic agent while despising its components, we might not trust the news we receive from Facebook, though we might like every individual person on it. Indeed, the anteater even talks about politics; the individual ants are communist in their political views, putting the common good before their own, but Aunt Hillary is a "laissez-faire libertarian".

The marvellous dialogue concludes with the sad story of the previous owner of the same anthill where Aunt Hillary lives. This was a very intellectual colony, which met with a tragic and untimely demise.

> *"One day, a very hot summer day, he was out soaking up the warmth, when a freak thundershower—the kind that hits only once every hundred years or so—appeared from out of the blue, and thoroughly*

drenched (him). Since the storm came utterly without warning, the ants got completely disoriented and confused. The intricate organisation which had been so finely built up over decades, all went down the drain in a matter of minutes. It was tragic".

But the ants did not die, they floated on sticks and logs, and when the waters receded, they managed to return to their home grounds. But there was no organisation left, *"and the ants themselves had no ability to reconstruct what had once before been such a finely tuned organisation".* Over the next few months, the ants which had been components of the previous owner *"slowly regrouped, and built up a new organisation. And thus was Aunt Hillary born".* But the two macroscopic agents, the anteater concludes, *"have nothing in common".*

This makes us wonder if the distinctive personality of each different social network might be reset if we were to restart them from scratch, like in a sort of ECT treatment. This might allow the very same participants to find a different organisation, and perhaps a healthier one.

In this perspective, we could even imagine echo chambers to be similar to a mental illness of this new kind of "mind", which could be cured somehow. Will there one day be a psychologist or a surgeon for large social machines that are too valuable to be turned off?

ON INVISIBLE HANDS AND MONKEY PAWS

The two strangers playing the ESP game may have no way of knowing that they are part of a teleological social machine, and even if they did, they would have no way of affecting its behaviour. This is also the case for the ants that compose Aunt Hillary: they just live at different levels of the social machine. And the same can be said for us, taking part in massive recommender machines whose ultimate goal is to attract and retain ever more participants.

Facebook, YouTube, and all the other entities that decide which content we should consume are socio-technical systems and have a form of intelligence which we can see in their capability to personalise their suggestions and to maximise their traffic. Similar insights apply to the reputation systems that maintain an estimate of the satisfaction that a user could expect from a service provider.

Their behaviour is not controlled by a homunculus within them but by the invisible hand that emerges from their internal dynamics. They have their own goals, which may be very different from those of their participants. As we interact with them, we also become part of them, shaping in small part their behaviour and being shaped by it, and we might not have a way to influence their ultimate direction of movement. Can we become trapped in them? What happens when their goals are in conflict with ours? Can these superhuman agents prove too powerful an adversary for us to resist?

If we do not know the answers to these questions, we should think twice before treating humans as elements of a machine, not only because this might harm their inherent dignity, but also because we might not be able to deal with the consequences.

In 1950, well before the web was invented and Tim Berners Lee coined the expression *social machine*, even before Douglas Hofstadter wrote the marvellous story of Aunt Hillary, the founder of cybernetics, Norbert Wiener, wrote a book about the relation between intelligent machines and society that ended with this prophecy, which refers back to the story of the monkey paw described in Chapter 5.

"I have spoken of machines, but not only of machines having brains of brass and thews of iron. When human atoms are knit into an organization in which they are used, not in their full right as responsible human beings, but as cogs and levers and rods, it matters little that their raw material is flesh and blood. What is used as an element in a machine, is in fact an element in the machine. Whether we entrust

our decisions to machines of metal or to those machines of flesh and blood which are bureaus and vast laboratories and armies and corporations, we shall never receive the right answers to our questions unless we ask the right questions. The Monkey's Paw of skin and bone is quite as deadly as anything cast out of steel and iron. The djinnee which is a unifying figure of speech for a whole corporation is just as fearsome as if it were a glorified conjuring trick. The hour is very late, and the choice of good and evil knocks at our door".

It is not possible to make predictions about which *unexpected* effects we might observe, but the near misses and other concerns of Chapter 5 and Chapter 7 can give us a starting point. I worry that we might one day discover that the monkey paw of W. W. Jacobs came in the form of an invisible hand.

10

REGULATING, NOT UNPLUGGING

We cannot realistically go back to a world without AI, so we need to find ways to live safely with it. Researchers are working on a list of principles that every agent will have to respect. This will need to be enforceable and verifiable, so the most fundamental property that we should expect of an agent is "auditability": being designed in a way that enables some form of inspection. It will then be possible to expect safety, fairness, privacy, transparency and other important requirements that legal scholars are debating at this time, so that we can trust intelligent machines. The shortcut that gave us the present form of AI makes this step difficult, but not impossible: replacing explicit rules with statistical patterns has made our machines difficult to read, and their decisions difficult to explain, but there can be different ways to inspect them, from enforcing internal checkpoints to some analogue of psychometrics for agents. Making

DOI: 10.1201/9781003335818-10

intelligent agents safe will require a much deeper understanding of the interface between social sciences, humanities, and natural sciences, and cannot be done by computer scientists alone. This is the next big challenge, and adventure, for AI.

CAN WE UNPLUG?

"You just have to have somebody close to the power cord. Right when you see it about to happen, you gotta yank that electricity out of the wall, man".

In October 2016, during an interview with Wired magazine, former US president Barack Obama joked on how to deal with *"AI's potential to outpace our ability to understand it"*. Like many jokes, it was funny because it was both absurd and revealing; how could we possibly turn off the intelligent agents that keep our essential infrastructure working?

This is a real problem; we currently rely on an infrastructure that in turn relies on AI, so that we get both costs and benefits of AI at the same time. Without spam filtering, fraud detection, and content recommendation, much of the web would not be able to function, and without personalised advertising, it might be unable to fund itself, unless there is a radical rethinking of its revenue model. Maybe not all of AI is in that delicate position, but it does seem to be the part that currently pays its bills. Computer scientists are used to thinking in terms of algorithms, but our current relation with AI is a socio-technical problem, where business models and legal and political issues interact.

Going back is not a realistic proposition, as we have phased out the previous infrastructures, which were anyway less efficient, and because anyone attempting this would have to face the competition of those who do not make the same choice; AI does increase productivity of the economy. This is the kind of dilemma in which many companies are currently trapped, but so are also all modern societies. The fact is that if we cannot live without AI, then we need to learn how to live with it on our terms.

The way forward is to face the problem of "trust", which includes understanding how we ended up in this situation and how this technology interacts with society so that we can manage that interaction. Not only do we need to understand the shortcuts that gave us the current version of AI so that we can try to imagine a different version, but we also need to understand the central position of control in which AI has been deployed. It is its location that enables AI both to make important decisions about us and to constantly observe our behaviour. And then we need to know what principles and rules we should put in place, so as to be sure that we do not regret the decades we spent developing this technology.

How did we end up, within a single generation, depending on a new technology that we are still learning to control? How can it affect our lives and what will we need to learn before we can fully trust it?

THE GREAT CONVERGENCE

On August 6th, 1991, Tim Berners Lee summarised his plans for the creation of a World Wide Web in a document called *"Short summary of the World Wide Web project"*, and then did something that might seem surprising: he posted it on the internet, inviting everyone to join.

How is that possible? The fact is that the pre-web world already used the internet, at least within the scientific or technical community, with telephone access to bulletin board systems and internet access to email and other information. What the WWW did was to seamlessly connect many different systems and enable easy access to them, so that it became possible for most "non-technical" people to join in. That started a chain reaction that is still under way.

By the time the dot-com bubble burst in 2000, only nine years after the first web servers were created, people were shopping, banking, reading news, and downloading music online. That included giving people computers, bringing them online, figuring out how to do secure payments, and format sound and video efficiently. This all happened in less than a decade; Bill Clinton was president when AOL first gave everybody commercial access to the web, and he was still there when the bubble burst.

Consider this: between 1994 and 1997, we witnessed the first fully online payment (secured by encryption), the creation of Amazon, eBay, Netflix, PayPal, and online banking, the first online newspapers (Daily Telegraph, Chicago Tribune, CNN), radio stations (WXYC, KJHK), and music streaming services (IUMA). At the same time, we also saw the creation of internet phone-calls and free web-based email services. And that was all before broadband or 3G; in 1998, the best connection users could have at home with dial up was 56Kbps, which also meant ringing an internet provider using your landline when you needed some information.

It was as if a sudden gravitational pull had started drawing together worlds that had always been separated. By the end of the 1990s, newspapers, radio stations, telephones, banks, payments, postal services, and telegraphs had all started to move to the internet thanks to the WWW browsers which integrated everything from encryption to media-playing software seamlessly, making it invisible to the user and able to run on home desktop computers. By the year 2001, ten years after Tim Berners-Lee's newsgroup post, 50% of Americans were online.

That also meant the end of a previous world; for each service that migrated online, there was a disruption in some sector of the economy and the ecosystems that surrounded it. Fax machines, aerial installers, news agents, and telegraph offices were among the first to be disrupted, but soon came the time for video rental stores, wire transfers, music shops, banks, cinemas, and then also radio, television, and ATMs. The parts of the previous infrastructure that survived were those which were co-opted and redefined by the new mass medium: for example, telephone cables and parcel delivery companies—and this process is still under way.

The number of webpages, services, and users grew so fast that very soon the web could not be managed without the help of intelligent algorithms, so a new sort of symbiosis was started. From 130 websites in 1993, the WWW went to about 250,000 in 1996 and 17 million in 2000. Search engines started being developed to direct users to relevant resources, starting an arms race with spammers that saw the emergence in the year 2000 of Google, who brought AI to that fight. Today, there are well over one billion websites, and there would be no hope in using the web without significant help from AI, both in locating resources and translating or summarising them, as well as in filtering spam and scams. The same applies to email, which would not be viable without very advanced spam filters. Frauds were also a problem, as were viruses, so while billions of people came to rely on the web for their lives, the web came to rely on AI for its own functioning.

The introduction of permanent connections accelerated the process; from roughly the year 2000, both broadband connections and Wi-Fi became available, users started being always connected, and from 2007 mobile phone connections were used too. Between 2001 and 2010, Apple launched the iPod, iTunes, the iPhone, and the iPad, eventually using 3G and 4G to connect to the web. This gave people the possibility of being constantly online, and today 85% of Americans and 86% of Europeans own a smartphone.

The burst of the dot-com bubble in 2000 created a new web, focused on revenue, and based on personalised advertising, competition for traffic, and user-generated content. From 2003 to 2007 we saw the creation of LinkedIn, Facebook, YouTube, Reddit, Twitter, Tumblr, and Instagram, some of which have billions of users and depend on AI at multiple levels, including indexing, retrieving, filtering, recommending, securing, and so on. Today, the monthly active users of Facebook are over three billion, those of YouTube around two billion, Instagram's about 500 million, those of TikTok around 50 million, Netflix has around 200 million, and Amazon has 300 million customers. Many of these services are related to the social machines that we discussed in Chapter 8.

Of course, AI is not limited to online applications, as there are many success stories of AI being used in healthcare or transportation. In 2018, a computer vision system called CheXNet was trained to detect pneumonia in X-ray images with a performance comparable to that of human experts, and as of 2022, autonomous cars have an average of ten accidents for every million miles of driving, which is not too much higher than what human drivers have. However, the largest stream of revenue for the AI field still comes from web-based applications.

The revolution started in the 1990s has permanently phased out an entire world, replacing it with a more efficient digital infrastructure where we conduct our business and our personal lives under the watchful eye of intelligent agents. Twenty Five years after Tim Berners-Lee published his project for a World Wide Web, Obama's joke about "pulling the plug" if AI ever goes out of control showed how irreversible that transition had become in only one generation.

MEDIUM AND GATEKEEPER

One of the great appeals of the web revolution was its potential to subvert existing power structures and shorten supply chains, allowing producers and consumers to bypass brokers and other intermediaries. This was the promise of dis-intermediation, directly connecting artists and audiences, sellers and buyers, jobseekers and employers.

In other words, one major appeal of this new medium was the possibility to bypass "gatekeepers": any actors who control access to opportunities or resources. In a few years, it became possible for anyone to produce a film and distribute it online, similarly for a book, a song, or a news article. Not only could this be done for free, with no friction and by ourselves, but in some cases this could also be done anonymously, which added to a feeling of liberation. Soon editors, publishers, producers, and journalists started being displaced. Employment, travel, and advertising agencies were next.

Many hailed dis-intermediation as a democratisation process, seeing intermediaries as gatekeepers, and a possible site where discrimination, exploitation, or corruption could occur, whereas the web promised a more "flatter" economy, without a middle-layer of services. And indeed, in many cases, it was possible to hold power to account thanks to the possibility of filming and directly sharing video. Some even called for disrupting the traditional style of representative democracy, replacing it with direct democracy where citizens can just vote for specific laws. After a decade of tumultuous convergence, the web was no longer displacing previous infrastructures and institutions, but also dis-intermediating a number of business models. By the early 2000s, after replacing previous communication infrastructures, the digital revolution was starting to reach cultural institutions such as the press, schools, and political parties.

The "moral panics" of 2016 highlighted a problem with that dream; it was intelligent algorithms, not people, matching job seekers with employers, artists with audiences, news articles with readers. And we were not sure if those decisions were made fairly, and if they could have unexpected effects on the wellbeing of the users. The media reports about possible gender and racial biases being absorbed "from the wild", and those about possible polarisation and circulation of fake news, highlighted how our technical understanding of those algorithms was limited and there was not the legal, cultural, and technical infrastructure to audit them and restore trust.

By 2016, not only did we start realising that we had new gatekeepers, but also that we could not trust them. And yet, they were in the position to observe the behaviour of billions of people, choose the information they could access, and make decisions about their lives. The power of intelligent agents derives from the position in which they are located. The main problem in AI today is that of trust.

LIVING WITH INTELLIGENT MACHINES

The purpose of automation is to replace people, and AI is no different. This can make individuals, organisations, and entire societies more productive, which is a way of saying that it can dramatically reduce the costs of certain tasks. On the one hand, this greatly empowers those who have access to that technology; on the other hand, it does create a problem for those workers that have been replaced or for those competitors who do not have access to the same resources. It is the wider economy, not only the digital one, that stands to benefit from AI, for example, by reducing the cost of access to transportation, translation, or medical diagnosis.

However, the same technology can also undermine some social values, for example by enabling mass surveillance via street cameras or mass persuasion via psychometric microtargeting. It can also cause harm, either when it malfunctions or when it causes unexpected side effects. It might even destabilise markets or public opinion and other collective goods, or accelerate the concentration of wealth in those who control the AIs or their data. It can be used for military applications in ways that we do not want to imagine.

Before the public can trust this technology, governments will have to regulate many aspects of it, and a lot can be learnt from the early stories of near misses and false alarms that have already emerged, as well as from the concerns expressed by the media or the stories of actual harm.

Two key factors will help to shape this landscape: accountability and auditability. Deciding who is accountable for the effects of an AI system will be a crucial step: is it the operator, the producer, or the user? And this links to the second factor: auditability. Are we going to trust systems which we cannot audit, perhaps even by design? Any further regulation of the sector will need to start from establishing that AI agents need to be auditable, and that burden falls on their producers or operators. Having established that, it will then be possible to discuss their safety, their fairness, and all other aspects which can only be established by inspecting the agents.

At this moment, many academic discussions on algorithmic discrimination have hit the obstacle that we do not actually know exactly how the COMPAS system scores defendants, how the Dutch benefits system was developed and used, or if any commercial tool for screening CVs has ever used the GloVe embeddings that we discussed (to remain within the examples of Chapter 5). We also do not know the contents of those studies that whistle-blower Frances Haugen released to the US authorities, claiming that they demonstrate some negative effects of Facebook products, such as excessive use and polarisation.

Making laws that are not based on experimental facts but only on media alarm is not a good idea, which means that we need to produce those studies on the effects of AI.

The obligation to be inherently auditable by design, so that a third party can ensure its safety, does not necessarily require disclosing code and should not be side-stepped by claiming that the internal models are inherently "not explainable", nor by replacing it with the usual legal disclaimers.

There may be some new science to be developed along the way, but we must expect that AI systems lend themselves to inspection, and their operators to account, though today this is not the case due to the technical shortcuts that we have taken to create AI. This might include the mandatory introduction of checkpoints within systems, the periodic run of standard stress-tests to assess their outcomes under extreme conditions, or even the development of a sort of "psychometrics" for machines, a science that does not yet exist.

Trust is not blind faith; it is a rational belief in the competence and the benevolence of an agent, whether that is a software, a person, or an organisation. Clear accountability is the way we ensure trust at this time.

On this basis, we should then address the remaining concerns: *safety, fairness, transparency, privacy, respect,* and so on. Any user should be able to expect the following from interactions with AI agents: safety (it will not cause harm, either by failures or by unintended side effects), respect (it will not attempt to nudge, manipulate, deceive, convince, or steer), transparency (it will declare its goals and motives, and the information it uses to pursue them), fairness (it will treat users equally when making decisions in key domains), and privacy (it will respect their right to control their personal data). In other words, the user should never have to worry about the motivations behind an agent's behaviour or any unexpected consequences of using it.

A hypothetical recommender system that causes addiction or other harm to minors, for example, would not be safe and should come with the option of being turned off or turned down. An autonomous car pilot that could cross red lights would also not be safe. A software agent that is constantly trying to manipulate the decisions of a user would not be respectful of their dignity and right to be left alone and should be expected to declare its actual goals, motives, and rewards. Even if many machine decisions are not explainable today, and even if the code has been released online, we should demand that a recommender engine declares its reward function publicly and plainly, as well as the signals it uses. Users should not have to wonder if a certain conversation or their location might be used as part of persuasive efforts, they should clearly know it from the start and openly have consented to it. Users should be able to opt out of parts of it without being penalised. A user uploading their CV on an online employment matchmaker should never have to worry about being discriminated against.

Trust is a multidimensional concept, and the requirements listed above are some of its dimensions and each of them is currently being investigated by scholars and policymakers around the world. The work that is required is not a matter of software engineering, but of understanding how the software interacts with society and individual psychology. That interface is where we will collectively decide which type of fairness we should expect and how to measure it as well as which kind of explanations we should expect from machines and how to assess them. There are many more dimensions to the problem of trust than the ones listed above, but there are also many clever proposals already on the table, and more are being developed, so there is every reason to be optimistic. All of that will be possible only if our agents are designed in a way to be auditable and AI companies are accountable for their services.

Since the same technology can cause concerns in one domain but not in another, and since not all levels of risk are the same, one idea is to regulate the applications of AI to specific problems and distinguish between high-risk or low-risk uses of AI; this is the direction currently taken by the European Union in its proposed regulations.

For example, in the current proposals, computing "social credit scores" of citizens by a state is listed as an "unacceptable risk" and would be prohibited. Even if this category is currently fairly empty, the very idea that there can be applications that are prohibited does set the right expectations. The "high-risk" category would come with many obligations of oversight and transparency and include any applications of AI that could result in loss of life (e.g. mass transport), access to education, employment, credit (sorting CVs, scoring exams, assessing loan applications), law enforcement, and judicial situations. It is not clear, however, how these would actually work in practice; would a CV-screening software need to be trained on hand-made high-quality data? Would data from the wild still be allowed? All of that will be the work of the next few years. The third tier of "limited risk" systems would mostly come with the obligation to disclose to the user that the agent is a machine, but I would also add the obligation of the agent to openly disclose its goals: is it rewarded to make you click or take any other actions?

Many questions remain unanswered by the current proposals: it is not clear to me where the concerns of addiction and polarisation deriving from recommender systems and personalised advertising would be addressed in the above taxonomy, and yet they are among the main concerns at this time. Remote biometrics would be banned by it (at least for commercial entities operating within the EU), but what about remote psychometrics? What level of anonymity should users have when consuming content and when publishing it? It will take a long time to learn what the actual risks of many AI applications are, and which remedies do work, but it is appropriate that we start regulating this space now.

The problem is that there are also risks in regulating this space without having all the experimental results, so we should expect rigorous studies in the short term on the effects of exposure to AI agents.

Other countries will come up with their own different proposals, but none will be enforceable if we accept that a software agent can be inherently beyond inspection and accountability by virtue of being opaque; auditability must be a precondition to be enforced. The burden of making the tools auditable should fall on the producer and a condition to be allowed to operate in the market. A system of licences might be unavoidable.

REGULATING, NOT UNPLUGGING

Obama's 2016 joke about dealing with the risks of AI outpacing our understanding, "*You just have to have somebody close to the power cord*", was funny and troubling for more than one reason. One is that we will not be able to unplug AI, as it has already become indispensable, and this is why we have to learn how to live safely with intelligent agents in all their various incarnations. The other is that it implies that we can decide when we have seen an intelligent agent crossing some line, but this might not be so easy, as we may have been looking for intelligence in the wrong place: AI is not coming in the form of a sentient robot, but rather of a learning infrastructure, perhaps a social machine, that makes crucial decisions for us and about us, with criteria that we cannot really understand. Its behaviour is shaped by statistical patterns found in human data and designed to pursue a goal. Regulating this kind of intelligent agent will be much more important than unplugging it, a project that will require significant work at the interface between natural, social, and human sciences. That will be the next cultural challenge for AI.

EPILOGUE

We have created a form of artificial intelligence (AI) and it is already part of our lives. It may not be what we had been expecting, but it does many of the things that we wanted, and some more, in a different way.

When we think of the intelligence in the intelligent agents that we created, it might be more useful to compare them with some distant and simple animal, such as the snails in our gardens, than to a person. In other words, their way of "thinking" is completely alien to us: they can learn, and some may even plan, but we cannot understand or argue with them, as they are driven only by patterns extracted from superhuman amounts of data, are only interested in pursuing their goal, and are indifferent to much else.

Nevertheless, they can be more powerful than us in some cases. We could worry at the realisation that this technology is now used in questionable ways, or we could marvel at the story of how cleverly we avoided a series of technical and philosophical questions along the way.

Don Antonio—the old priest who tried to quiz my first computer on Greek history—was not interested in technical details but in the point of view of people. He might not have cared too much about how Siri can now answer his question about Alexander the Great, an explanation that would involve an enormous list of events that happened after his death, from Wikipedia to smartphones. But he would have probably detected some disturbing shift in our language, as we explain all that happened: why do we call people "users" and the expressions of their art and culture "content"? Who would ever refer to our town's great wines as "content"? Maybe just a bottle merchant.

I do not think he would have minded that machines can be intelligent too, maybe not even that they can be more intelligent than people at certain practical tasks. But he would have definitely objected to them being as important, or more important, than people. Adopting their viewpoint and language as we describe the new world we are creating may be an alarming step.

The old analog world that he inhabited is now gone, so we will need to adapt our culture, but it has not yet entirely disappeared from our memory. A child can still see in her smartphone the icons of an envelope, a telephone handset, a camera with a lens and shutter button, a printed newspaper, and an analog clock, each representing a different app. Some of today's children have never handled those objects, or even used a phone box, and one day this might be the case also for cash. I doubt that many children know the meaning of Tube and Flix, and there have been reports of young children attempting to scroll the pages of a book with their fingers.

The magic device in their pockets acts at once as a camera, a telephone, a mailbox, a television, a newspaper, a bank card, and so on. That is a fantastic medium that is transforming entire communities in poor countries, and empowering individuals in all sorts of ways; some think it is responsible for revolutions and migrations, and certainly it is responsible for new types of art and beauty.

That is also where our children meet AI: it is AI that answers their questions, recommends music and news, translates across languages, and retrieves and filters information. We can look forward to much more: doctors will soon use it to diagnose certain diseases or access updated information about rare conditions. Leaving the phone aside, AI will soon be found in hospitals, cars, and schools. Not only would it be impossible to go back, it would be irresponsible: what we have to do instead is make this technology safe.

This is where Don Antonio would have something to say. We have created a world for our children where the machines will make decisions for them and about them. We owe it to the new generation, who carry those machines in their pockets and have no chance of understanding what we have left behind, that they can trust the world we created for them. They will upload their CV, college admission statement, or whatever other important request and then will wait for a decision by the agent. They should be certain that these machines will not exploit them or their dreams, nor discriminate against them, nor let them down in any other way.

There will be pressure to look the other way, and we should prepare for that too. When the computer drives more safely than a person, some producers will argue, is it ethical to insist on driving your car? When the software can make better predictions about a loan or a job application, is it acceptable to opt out or request to use humans instead? We may also soon be called to decide what to do with our values when our economic competitors violate them or when there is an emergency. We will hear these objections very soon, so let us start thinking about them now.

Our cultures will evolve, somehow, to incorporate this new presence. What we should keep on teaching to the next generation, however, is that the highest value is the dignity of human beings, and this is how we should measure every future decision involving the role of intelligent machines. Regardless of how much cleverer than us they will become, "they will never be better than us". Perhaps Don Antonio would have appreciated the defiant classified ad posted by those Amazon editors on a magazine in Seattle and directed at Amabot, "*the glorious messiness of flesh and blood will prevail*". I really think that it will.

Bibliography

CHAPTER 1

Albus, James S. "Outline for a Theory of Intelligence." *IEEE Transactions on Systems, Man, and Cybernetics* 21.3 (1991): 473–509.

Sagan, Carl. "The Planets", *Christmas Lectures*. The Royal Institution, 1977, BBC2.

Wiener, Norbert. *Cybernetics: Or Control and Communication in the Animal and the Machine*. Paris: John Wiley and Sons/ Technology Press, 1948 (2nd revised ed. 1961).

CHAPTER 2

Boyan, Justin, Dayne Freitag, and Thorsten Joachims. "A machine learning architecture for optimizing web search engines." *AAAI Workshop on Internet Based Information Systems*. 1996.

Cristianini, N. "Shortcuts to Artificial Intelligence." In Marcello Pelillo and Teresa Scantamburlo (eds), *Machines We Trust*. Cambridge, MA: MIT Press, 2021.

Goldberg, David, David Nichols, Brian M. Oki, and Douglas Terry. "Using Collaborative Filtering to Weave an Information Tapestry." *Communications of the ACM* 35.12 (1992): 61–70.

Halevy, Alon, Peter Norvig, and Fernando Pereira. "The Unreasonable Effectiveness of Data." *IEEE Intelligent Systems* 24.2 (2009): 8–12.

McCarthy, John, Marvin L. Minsky, Nathaniel Rochester, and Claude E. Shannon. "A Proposal for the Dartmouth Summer Research Project on Artificial Intelligence, August 31, 1955." *AI Magazine* 27.4 (2006): 12–12.

Vapnik, Vladimir. *The Nature of Statistical Learning Theory.* New York: Springer Science & Business Media, 1999.

CHAPTER 3

"A Compendium of Curious Coincidences." *Time* 84.8 (August 21, 1964).

Borges, Jorge Luis. "The Analytical Language of John Wilkins." *Other Inquisitions* 1952 (1937): 101–105.

CHAPTER 4

Lovelace, Augusta Ada. "Sketch of the Analytical Engine Invented by Charles Babbage, by LF Menabrea, Officer of the Military Engineers, with Notes Upon the Memoir by the Translator." *Taylor's Scientific Memoirs* 3 (1842): 666–731.

Samuel, A. L. "Some Studies in Machine Learning Using the Game of Checkers." *IBM Journal of Research and Development* 3.3 (1959): 210–229. doi: 10.1147/rd.33.0210.

Silver, David, Aja Huang, Chris J. Maddison, Arthur Guez, Laurent Sifre, George van den Driessche, Julian Schrittwieser, Ioannis Antonoglou, Veda Panneershelvam, Marc Lanctot, Sander Dieleman, Dominik Grewe, John Nham, Nal Kalchbrenner, Ilya Sutskever, Timothy Lillicrap, Madeleine Leach, Koray Kavukcuoglu, Thore Graepel, and Demis Hassabis. "Mastering the Game of Go with Deep Neural Networks and Tree Search." *Nature* 529.7587 (2016): 484–489.

CHAPTER 5

Angwin, Julia, Jeff Larson, Surya Mattu, and Lauren Kirchner. "Machine Bias." *Propublica*, May 23, 2016. https://www.propublica.org/article/machine-bias-risk-assessments-in-criminal-sentencing.

"Boete Belastingdienst Voor Discriminerende en Onrechtmatige Werkwijze." December 7, 2021. https://www.autoriteitpersoonsgegevens.nl/nl/nieuws/boete-belastingdienst-voor-discriminerende-en-onrechtmatige-werkwijze.

"Cabinet Admits Institutional Racism at Tax Office, But Says Policy Not to Blame." *nltimes.nl*, May 30, 2022. https://nltimes.nl/2022/05/30/cabinet-admits-institutional-racism-tax-office-says-policy-blame.

Caliskan, A., J. J. Bryson, and A. Narayanan. "Semantics Derived Automatically from Language Corpora Contain Human-like Biases." *Science* 356.6334 (2017): 183–186.

Dastin, Jeffrey. "Amazon Scraps Secret AI Recruiting Tool That Showed Bias against Women." *Reuters*, October 11, 2018.

Hadwick, D., and S. Lan. "Lessons to Be Learned from the Dutch Childcare Allowance Scandal: A Comparative Review of Algorithmic Governance by Tax Administrations in the Netherlands, France and Germany." *World Tax Journal* 13.4 (2021).

Jacobs, W. W. "The Monkey's Paw (1902)." In W. W. Jacobs (ed.), *The Lady of the Barge*. London and New York: Harper & Brothers, 1902.

Kleinberg, Jon, Sendhil Mullainathan, and Manish Raghavan. "Inherent Trade-Offs in the Fair Determination of Risk Scores." 2016. *arXiv preprint arXiv:1609.05807*.

"PwC Onderzoekt Doofpot Rondom Memo Toeslagen Affaire." June 14, 2021. https://www.accountant.nl/nieuws/2021/6/pwc-onderzoekt-doofpot-rondom-memo-toeslagenaffaire/.

Rudin, Cynthia, Caroline Wang, and Beau Coker. "The Age of Secrecy and Unfairness in Recidivism Prediction." 2018. *arXiv preprint arXiv:1811.00731*.

"Tax Authority Fined €3.7 Million for Violating Privacy Law with Fraud Blacklist." *nltimes.nl*, April 12, 2022. https://nltimes.nl/2022/04/12/tax-authority-fined-eu37-million-violating-privacy-law-fraud-blacklist.

Wiener, Norbert. *The Human Use of Human Beings: Cybernetics and Society*. Boston: Houghton Mifflin Publisher, Revised 2nd ed., 1954.

CHAPTER 6

Burr, C., and N. Cristianini. "Can Machines Read Our Minds?" *Minds and Machines* 29.3 (2019): 461–494.

Duhigg, C. "How Companies Learn Your Secrets." *The New York Times* 16.2 (2012): 1–16.

Eichstaedt, Johannes C., Robert J. Smith, Raina M. Merchant, Lyle H. Ungar, Patrick Crutchley, Daniel Preoţiuc-Pietro, David A. Asch, and H. Andrew Schwartz. "Facebook Language Predicts Depression in Medical Records." *Proceedings of the National Academy of Sciences* 115.44 (2018): 11203–11208.

Graff, Garrett. "Predicting the Vote: Pollsters Identify Tiny Voting Blocs." *Wired Magazine*, 16, 2008.

Kosinski, M., D. Stillwell, and T. Graepel. "Private Traits and Attributes Are Predictable from Digital Records of Human Behaviour." *Proceedings of the National Academy of Ssciences* 110.15 (2013): 5802–5805.

"Letter to the DCMS Select Committee by Information Commissioner Office." October 2, 2020. https://ico.org.uk/media/action-weve-taken/2618383/20201002_ico-o-ed-l-rtl-0181_to-julian-knight-mp.pdf.

Matz, S. C., M. Kosinski, G. Nave, and D. J. Stillwell. "Psychological Targeting as an Effective Approach to Digital Mass Persuasion." *Proceedings of the National Academy of Sciences* 114.48 (2017): 12714–12719.

Nix, Alexander. "From Mad Men to Math Men." March 3, 2017. https://www.youtube.com/watch?v=6bG5ps5KdDo.

CHAPTER 7

Allcott, H., L. Braghieri, S. Eichmeyer, and M. Gentzkow. "The Welfare Effects of Social Media." *American Economic Review* 110.3 (2020): 629–676. https://www.aeaweb.org/articles?id=10.1257/aer.20190658.

Auxier, B., and M. Anderson. "Social Media Use in 2021." *Pew Research Center* 1 (2021): 1–4.

Burr, C., N. Cristianini, and J. Ladyman. "An Analysis of the Interaction between Intelligent Software Agents and Human Users." *Minds and Machines* 28.4 (2018): 735–774.

Cheng, Justin, Moira Burke, and Elena Goetz Davis. "Understanding Perceptions of Problematic Facebook Use: When People Experience Negative Life Impact and a Lack of Control." *Proceedings of the 2019 CHI Conference on Human Factors in Computing Systems*, 2019. https://dl.acm.org/doi/10.1145/3290605.3300429.

D'Onfro, J. "The 'Terrifying' Moment in 2012 When YouTube Changes Its Entire Philosophy." *Business Insider*, 2015. https://www.businessinsider.com/youtube-watch-time-vs-views-2015-7.

Eichstaedt, Johannes C., Robert J. Smith, Raina M. Merchant, Lyle H. Ungar, Patrick Crutchley, Daniel Preoţiuc-Pietro, David A. Asch, and H. Andrew Schwartz. "Facebook language predicts depression in medical records." *Proceedings of the National Academy of Sciences* 115, no. 44 (2018): 11203–11208.

Kramer, Adam D. I., Jamie E. Guillory, and Jeffrey T. Hancock. "Experimental Evidence of Massive-scale Emotional Contagion through Social Networks." *Proceedings of the National Academy of Sciences* 111.24 (2014): 8788–8790.

Levy, R. E. "Social Media, News Consumption, and Polarisation: Evidence from a Field Experiment." *American Economic Review* 111.3 (2021): 831–870.

Newton, C. "How YouTube Perfected the Feed." *The Verge* 30 (2017).

Ofcom, U. K. *Children and Parents: Media Use and Attitudes Report*. London: Office of Communications, 2021.

Ortiz-Ospina, Esteban. "The Rise of Social Media." *Our World in Data* 18 (2019).

"The Facebook Files." *Wall Street Journal*, 2021. https://www.wsj.com/articles/the-facebook-files-11631713039.

Walker, Andrew. Coroner's Service; Molly Russell: Prevention of future deaths report (Reference No. 2022-0315). North London (October 14, 2022). Retrieved from: https://www.judiciary.uk/prevention-of-future-death-reports/molly-russell-prevention-of-future-deaths-report/.

CHAPTER 8

Badia, A. P., B. Piot, S. Kapturowski, P. Sprechmann, A. Vitvitskyi, Z. D. Guo, and C. Blundell. "Agent57: Outperforming the Atari Human Benchmark." In *International Conference on Machine Learning* (pp. 507–517). PMLR, 2020.

Bellemare, Marc G., Y. Naddaf, J. Veness, and M. Bowling. "The Arcade Learning Environment: An Evaluation Platform for General Agents." *Journal of Artificial Intelligence Research* 47 (2013): 253–279.

Chrabaszcz, P., I. Loshchilov, and F. Hutter. "Back to Basics: Benchmarking Canonical Evolution Strategies for Playing Atari." 2018. *arXiv preprint arXiv:1802.08842*.

Ie, E., Vihan Jain, Jing Wang, Sanmit Narvekar, Ritesh Agarwal, Rui Wu, Heng-Tze Cheng, Morgane Lustman, Vince Gatto, Paul Covington, Jim McFadden, Tushar Chandra, Craig Boutilier. "Reinforcement Learning for Slate-based Recommender Systems: A Tractable Decomposition and Practical Methodology." 2019. *arXiv preprint arXiv:1905.12767*.

CHAPTER 9

Berners-Lee, Tim, and Mark Fischetti. *Weaving the Web: The Original Design and Ultimate Destiny of the World Wide Web by Its Inventor*. San Francisco: Harper, 1999.

Cristianini, Nello, and Teresa Scantamburlo. "On Social Machines for Algorithmic Regulation." *AI & Society* 35.3 (2020): 645–662.

Cristianini, Nello, Teresa Scantamburlo, and James Ladyman. "The Social Turn of Artificial Intelligence." *AI & Society* (2021): 1–8.

Goldberg, David, David Nichols, Brian M. Oki, and Douglas Terry. "Using Collaborative Filtering to Weave an Information Tapestry." *Communications of the ACM* 35.12 (1992): 61–70.

Hofstadter, Douglas R. *Gödel, Escher, Bach*. New York: Basic Books, 1979.

O'Reilly, T. "Open Data and Algorithmic Regulation." In B. Goldstein and L. Dyson (eds), *Beyond Transparency: Open Data and the Future of Civic Innovation* (pp. 289–300). San Francisco: Code for America Press, 2013.

Smith, Adam. *An Inquiry into the Nature and Causes of the Wealth of Nations: Volume One*. London: W. Strahan and T. Cadell, 1776.

Von Ahn, Luis. "Games with a Purpose." *Computer* 39.6 (2006): 92–94.

Wiener, Norbert. *The Human Use of Human Beings: Cybernetics and Society*. Houghton Mifflin Publisher, Revised 2nd ed., 1954.

CHAPTER 10

Dadich, Scott. "The President in Conversation with MIT's Joi Ito and WIRED's Scott Dadich." 2017. https://www.wired.com/2016/10/president-obama-mit-joi-ito-interview/.

INDEX

A

agent, 4–12, 13, 25–33, 35, 40, 43, 50, 55–56, 58, 64, 66, 71–72, 74, 80, 84, 98–105, 107, 113, 116–117, 120–127, 133, 135–140, 144, 149, 153, 155–161, 163
Albus, James, 5
AlphaGO, 53–54, 56, 57, 59–61, 71, 84, 125
Amabot, 25, 28, 116
Amazon, 25–28, 29, 30, 31–32, 36, 69, 116, 151, 153, 164
Antonio, Don, ix, x, 162, 163, 164
Aplysia, 38–39, 40, 42, 50
Arcade Learning Environment (ALE), 120–121
Aristotle, 138
artificial intelligence (AI), 3–4, 10, 11, 14, 19, 20–25, 30–32, 33, 35, 50, 53, 57, 59, 61, 64, 66, 71, 73, 79, 84, 93, 95–96, 98, 120–121, 125, 130, 134, 149–150, 152–160, 161, 163
Ashby, Ross, 60
Atari, 119, 120, 126

B

Babbage, Charles, 51, 52
Beer, Stafford, 131
Berg, Howard, 49
Berners-Lee, Tim, 135, 146, 150, 151, 153
Bezos, Jeff, 26, 116
Big Five, 81, 83
Borges, Jose Luis, 46–47, 60, 68, 74
Breiman, Leo, 35

C

Caliskan, Aylin, 72
Cambridge Analytica, 77, 91, 92, 94, 96
checkers, 57–58
collaborative filtering, 27, 32, 139–140

concepts, 2, 9, 25, 40, 46, 60, 72, 82, 101, 136, 157
constructs, 45–48, 68, 74, 80, 81
cybernetics, 5, 63, 75, 134, 136, 146

D
data brokers, 91–93, 95
Davis, Warren, 119
DeepMind, 53
Deep Q-Network (DQN), 121–123

E
Einstein, Albert, 37
Empirical Inference, 37
ESP game, 128–130, 132, 133, 136, 145
expert systems, 21–22, 24, 31

F
Facebook, 66, 67, 80, 82–83, 85–86, 88–89, 94, 108–112
Feigenbaum, Ed, 21, 28, 33
Fen Hui, 53
Feynman, Richard, 99, 116

G
Games with a Purpose (GWAP), 130–132
Go, 53, 55, 57, 59, 126
Goldberg, Dave, 27, 32, 36, 139–140
Google, 30, 33, 49, 53, 71, 103, 130, 152
GPT-3, 36, 49, 61, 64, 71, 84

H
Haugeland, John, 21
Haugen, Francis, 109, 111, 156
Hebb, Donald, 9
Hillary, Aunt, 143–146
Hofstadter, Douglas, 143, 146
Hume, David, 39

I
intelligence, 1–7, 9, 10–15, 20, 24, 35, 37, 57, 61, 99, 138, 140, 146, 160, 161

invisible hand, 133, 138, 140, 145–147

J
Jacobs, W.W., 63, 75, 147
Jelinek, Frederick, 16–20, 22, 23, 24, 27, 28, 31, 33, 37, 49, 50, 72, 96

K
Kasparov, Gary, 54
Kennedy, John F., 44
Kuhn, Thomas, 29

L
LaMDA, 49
Lee, Jeff, 119
Lévi-Strauss, Claude, 61
Lincoln, Abraham, 44
Linden, Gary, 27
LISP machines, 21, 24
Lovelace, Ada, 51–54, 56, 60, 61

M
machine learning, 19, 24–25, 28, 31, 34, 35, 40–42, 45, 48–49, 53, 54–58, 64, 72, 74, 83, 90, 100–101, 107, 120
Mayer, Marissa, 103
McCarthy, John, 11, 20
mechanism design, 132
Menabrea, Luigi, 51, 52, 61
Mercer, Robert, 96
monkey paw, 62–63, 73–75, 145–147
Mott, Bradford, 120

N
Nix, Alexander, 76–79, 80, 84, 87, 93, 95, 96
nudge, 88, 93, 102–103, 106–107, 108, 112, 116, 117, 124, 125–126, 157

O
Obama, Barack, 92, 149, 153, 159
O'Reilly, Tim, 141

P
paradigm shift, 29–30
Pascal's triangle, 40, 42
patterns, 4, 10, 12, 18, 19, 25, 28, 29, 31, 34, 36, 41–45, 47–48, 50, 66, 69, 74, 81–82, 100, 160, 161
Polya, George, 3
POSIWID principle, 131
psychometrics, 80–84, 156, 159

Q
Q*bert, 119–120, 121, 122, 126–127

R
Russell, Bertrand, 39, 113

S
Sagan, Carl, 2–3, 10, 14
Samuel, Arthur, 57–59
Samuelson, Paul, 103
Sedol, Lee, 53, 59

shortcut, 32, 37, 64, 66, 68–69, 73, 89–90, 96, 123, 124, 150, 156
Smith, Adam, 133, 138
social machines, 134–137
Stella, 120

T
Trump, Donald, 77–78
Turing, Alan, 10, 52, 135
Turing's Test, 4

V
Vapnik, Vladimir, 34, 37, 49
von Ahn, Luis, 130, 131

W
Wiener, Norbert, 63, 75, 134, 146
Wigner, Eugene, 30–31
Wittgenstein, Ludwig, 50

Y
Yeung, Karen, 107

Milton Keynes UK
Ingram Content Group UK Ltd.
UKHW022050141024
449569UK00031B/1573

9 781032 305097